Practical Manual on
WEED MANAGEMENT

THE AUTHORS

Dr. Vivek is a well known professor of Agronomy and he has professional experience of more 30 years of Teaching, Research and Extension with specialization of Weed and Nutrient Management at Sardar Vallabhbhai Patel University of Agriculture & Technology, Meerut-250110(U.P.), India. He has guided more than 12 Ph.D. and 15 M.Sc. students in their research programme. He is prolific writer and has published several manuals for teaching different courses, a lot of chapters in scientific books, more than 125 research papers published in refereed national and international Journals and many popular articles in different rural development magazine for the welfare of farmers. Dr. Vivek has also authored books (Agriculture and Climate Change etc). He has chaired a number of scientific sessions in different symposium, conferences and workshops and made recommendation to government for adoption and implementation. He is regularly called as resource person to deliver invited talks in seminars and symposium and conferences. He acted as member of organizing committee of many seminar and symposiums. He is life member of 06 professional societies.

Mausmi Rastogi is currently pursuing Ph.D in Agronomy from Sardar Vallabhbhai Patel University of Agriculture & Technology Modipuram Meerut -250110(U.P.), India. She did her B.Sc. (Ag.) from Chaudhary Charan Singh University in 2019 and M.Sc. in Agronomy from Sardar Vallabhbhai Patel University of Agriculture & Technology Modipuram, Meerut, Uttar Pradesh with 1st division in year 2021. She has awarded with "Best Research Scholar Award in 2022" and has published many research papers, articles, and book chapters in many national and international peer reviewed journals and also 2 book. She has attended many conferences on National and International level.

Gajjela Indira is pursuing Ph.D in Agronomy from Sardar Vallabhbhai Patel University of Agriculture & Technology Modipuram Meerut -250110 (U.P.), India. Also M.Sc. in Agronomy from same university and secured first position from her batch. She obtained her B.Sc. agriculture degree from Agriculture College, Bapatla, Acharya N. G. Ranga Agricultural University, Andhra Pradesh, India. and was adjudged with the Best student award (2015-16).She has published many research papers, articles, and book chapters in many national and international peer-reviewed journals.

Shikhar Verma is currently pursuing Ph.D in Agronomy from Chandra Shekhar Azad University of Agriculture & Technology Nawabganj, Kanpur, 208002 (U.P.), India . He did his B.Sc. (Ag.) Hons. from Lucknow University in 2019 and M.Sc. in Agronomy from Sardar Vallabhbhai Patel University of Agriculture & Technology Modipuram, Meerut, Uttar Pradesh with Ist division. He has awarded with "Best Research Scholar Award in 2022" and has published many research papers, articles, and book chapters in many national and international peer-reviewed journals and also Published 3 books.

Sushmita obtained her bachelor's in agriculture from University of Agricultural Sciences, Bengaluru and Masters in (Agriculture) Agronomy from University of Agricultural Sciences, Dharwad. She has cleared "ICAR NET" in Agronomy and "UGC NET" in Environmental sciences. She is currently a Ph.D. research scholar in Sardar Vallabhbhai Patel University of Agriculture and Technology, Meerut India.

Practical Manual on
WEED MANAGEMENT

Dr. Vivek
Professor Agronomy,
Sardar Vallabhbhai Patel University of Agriculture & Technology Modipuram
Meerut, Uttar Pradesh, India.

Mausmi Rastogi
Research Scholar,Department of Agronomy,
Sardar Vallabhbhai Patel University of Agriculture & Technology Modipuram
Meerut, Uttar Pradesh, India.

Gajjela Indira
Research Scholar, Department of Agronomy,
Sardar Vallabhbhai Patel University of Agriculture & Technology Modipuram
Meerut, Uttar Pradesh, India.

Shikhar Verma
Research Scholar, Department of Agronomy,
Chandra Shekhar Azad University of Agriculture & Technology Nawabganj,
Kanpur, Uttar Pradesh, India.

Sushmita
Research Scholar, Department of Agronomy,
Sardar Vallabhbhai Patel University of Agriculture & Technology Modipuram
Meerut, Uttar Pradesh, India.

2023

Daya Publishing House®

A Division of

Astral International Pvt. Ltd.

New Delhi – 110 002

Cataloging in Publication Data--DK

Courtesy: D.K. Agencies (P) Ltd. <docinfo@dkagencies.com>

Vivek (Professor of agronomy), author.

Practical manual on weed management / Dr. Vivek, Mausmi Rastogi, Gajjela Indira, Shikhar Verma, Sushmita.

pages cm

ISBN 9789359191683

1. Weeds--Control. I. Title.

LCC SB611.V58 2023 | DDC 632.5 23

Published by : **Daya Publishing House**®
A Division of
Astral International Pvt. Ltd.
– ISO 9001:2015 Certified Company –
4736/23, Ansari Road, Darya Ganj,
New Delhi-110 002
Ph. 011-43549197, 23278134
E-mail: info@astralint.com
Website: www.astralint.com

Preface

Weeds are plants that grow in places where other, desirable plants should grow. It is true that a plant may be a weed in some places but not in others. Each weed has its way of winning its struggle with the crops and hence its habit must be learned to know how to get rid off weeds. This can be done only by studying the life history of weed species. Different ways must be used to get rid of weeds based on how they grow and what they are made of. This manual is designed to cover all of the practical aspects of weed management, from collecting weeds to figuring out how they look. This will help you figure out how to get rid of different kinds of weeds in different kinds of fields. The manual also has useful information about how to use herbicides in the field, including how to apply them, calibrate spray pumps, and do different kinds of calculations. Photographs, practical exercises, and real-life examples back up this information. A study of the ITK's that people in the area used to get rid of weeds has also been planned. The different things this manual talks about will help students at the undergraduate and postgraduate levels learn about weeds and how to get rid of them. It will also help teachers, extension workers, and scientists who work in the field of weed management, learn the latest technical information about weed ecology in the field. The weed manual will also be helpful for people who want to learn more about weed management and for people who need to help farmers or other workers become better at weed management. In several cases, it has not been possible to obtain period permission for reproduction for which the authors & publishers offer their sincere apologies. The goal of this manual is to cover all practical aspects of weed management based on a unified syllabus set by the Indian Council of Agricultural Research (ICAR).

Authors

Contents

Exercise -1

Collection of Weed Specimens and Preparation of Herbarium

Collection of weeds is essential to know about the morphology and biology of weeds, so as to decide effective control of weeds under field conditions.

Herbarium is a collection of plants that have been dried, pressed, and storedon herbarium sheets according to a particular purpose for future reference, record, and study. Because a herbarium is a repository of information such as geographical ranges, taxonomic, biological, and ecological data, each specimen represents a voucher deposited for future use. Collecting weeds and storing them in a herbarium is an essential aspect of weed science education, research, and extension.

Objectives

a) To get familiar with thc common weeds.
b) To learn the art of collection of weed specimen and preparation of herbarium.
c) To study the morphology and growth habit of weeds.

Materials

Herbarium sheets, wooden block press, old newspaper, pencil or marker, thread.

Procedure

The scientific method for collection of weed specimen consists of the following important steps:

a) Collection of plant sample from field.
b) Pressing and drying of collected specimen.
c) Mounting on herbarium sheet.
d) Fixing of identification label.

a) Collection of Plant Sample from Field

Healthy and ideal plants which are free from insect-pests and diseases are selected. Weed specimen should contain all parts of plant including root, stem, leaves, flowers fruits etc. For good identification, no part of the plant should be excluded. The collected plants are placed in polythene bags separately to avoid mixing of species. Two to three specimens may be collected to select the best one. Every species should be given identification number. If plants are too small or large, extra care should be taken while collecting the specimen. A large weed plant may be divided into 2, 3 or more sections and each section should be pressed separately. However, excess branches or leaves may be removed provided remaining leaves and branches truly represent the plant. If the plants are very small, more number of specimens of the same plant should be collected.

b) Pressing and Drying of Collected Specimen

Once the weed specimen is rooted out at right stage (at seedling, flowering and fruiting), it is necessary to press it and fit well inside the limits of folded sheets of paper. Weed with long stem or leaves may be folded into V, N or W bends but should not be doubled back in such a way as to lie across itself. If the specimen can't be directly dried they have to be preserved by preservatives. If the foliage is very thick, it has to be pruned assuring that the portion of the cut parts are identified. If the leaf is large, cut it vertically into 2 halves use one of the portions as specimen. The newspaper/bolting paper changed after 12 hrs in first incidence and these after 24 hrs, 48 hrs and 72 hrs is done till the specimen is dried completely. This is called **Natural drying**. In **artificial drying**, after sweating period specimens are direct in hot air oven by maintaining 62^{0}C.

c) Mounting on Herbarium Sheet

Well pressed weed sample should be mounted on the herbarium sheet with the help of cellophane tape at the centre of the sheet.

d) Fixing of Identification Label

The identification number should accompany throughout every stage of herbarium preparation process. It consists of following two steps:

(I) Collection of Information

The person who gets a sample of a weed should write down as much useful information as possible at the time of the collection. The information can be written on the edge of the newspaper that will be used to bring the weed sample from the field to the lab, or it can be written in a diary with a reference number for each weed sample. On sunny days, collection should be done in the afternoon. The collector should keep track of information about the following things.

i. **Location:** Name of the village or town nearby the field and its distance and direction from the known town. For exact location, the district may be useful. For example, 10 km east from Meerut city, Road side.

ii. **Date:** The data should be clearly mentioned with day, month and year. It should be written as December 27, 2022.

iii. **Habitat:** Under this category, name of place with ecological conditions of site viz., field, pasture, roadside, woods, hillside, sanddune, nullah (eroded stream), light exposure (sun or shade), moisture conditions (dry, moist, wet etc.) and denseness of community (bare ground, thin or dense population) should be pointed out.

iv. **Occurrence of Weed:** A weed under consideration should be described in relativity of number of other species of weeds. For this purpose, an arbitrary scale for comparison using terms like rare, occasional, frequent, common and abundant may be followed.

v. **Noting of Essential Characteristics:**

   Nature-Annual, biennial, perennial

   Root-Tap, fibrous, adventitious, shallow, deep Stem/branches - Woody, herbaceous, erect, spreading, trailing, prostrate

   Leaves - Simple, compound, heart shaped, narrow, board Flower - Shape, colour, fragrance, pollination

vi. **Features of Special Reference:** Some plants in nature are known for their special characteristics in terms of fragrance, colour, leaf curling, stringing hairs, double colour of leaves, milky juice of stem or leaves, habitat of growth, stickiness etc. The specific characteristics of plants along with right stage of growth and development of the plant should be mentioned.

vii. **Miscellaneous Points of Interest:** A collector by his own Wisdom or discussion with local people may collect valuable and rare information about a weed specimen. This includes special use, preference shown by insect-pest, industry etc. special control measure, anything special about dissemination and propagation etc.

(II) Format of Identification Label

A scientific identification label should include the following points

State	:	**Ref. No.**
District	:	
Location and Habitat	:	
Common Name (English)	:	
(Local)	:	
Scientific Name	:	
Description	:	
Collectors Address	:	
Date & Time	:	
Date & Time	:	

Example:

LABEL		**Ref. No. 1**
State	:	Uttar Pradesh
District	:	Meerut
Location and Habitat	:	10 km eastfrom Meerut city, Road side
Common Name (English)	:	Field bind weed
(Local)	:	Hiran Khuri
Scientific Name	:	Convolvulus arvensis
Description	:	Perennial herb, broad leaf, trailing stem creeping roots, loamy sand, wheat field
Collectors Address	:	Mausmi Rastogi, B.Sc. (Ag.) Ist year
Date & Time	:	December 27, 2022

Exercise - 2

Weed Identification

Correct identification is an important step in working out the strategy to control weeds. Knowledge of morphology is most important in the identification of weeds. There are approximately 250,000 species of plants in the world. However, less than 250 plant species, about 0.1% are troublesome enough to be called universally throughout the world.

Materials

Manuals and books on weed management, Botany, Taxonomy, Weed Science, Herbaria, HB Pencil, White paper etc.

Procedure

Look at the weed samples that you saved in the last exercise and try to figure out what their main traits are. After you find where different kinds of weeds grow again, you can study their habitat, shape, and how they spread. Use books, manuals, herbaria, exhibits, the Internet, and other sources to report on weed specimens in a way that is correct and scientific. Some of the weeds are shown on plates in this manual.

Important scientific information may be tabulated as under:

Table 2.1: Identification of Weeds

S. No.	Name of weed (Local/English)	Botanical name	Group	Family	Salient characteristics
1	Bermuda grass	Cynodon dactylon	Narrow leaf	Poaceae	Perennial, ventitious roots, stolon

Table 2.2: Important Preserved Weeds in the Album

S. No.	Common Name (English/ Local	Scientific Name	Group	Family	Salient Characteristics
1.					
2.					
3.					
4.					
5.					
6.					

Common Weeds

Brachaiaria Repatans

Cenchrus Ciliaris

Cenchrus Longispinosus

Dactyloctenium Aegypticum

Digitaria Sanguinalis

Eleusine Indica

Echinochloa Colona

Echinocnloa Glabrescens

Leptochloa Chinensis

Ischarmum Rugosum

Paspalum Distichum

Sorghum Halepense

Paspalum Conjugatum

Digitaria Sanguinalis

Panicum Dichotomiflorum

Amaranthus Viridis

Commelina sp

Aeschynomene Indica

Ammannia Baccifera

Monochoria Vaginalis

Cyperus Iria

Eliocharis

Bulbostylis Barbata

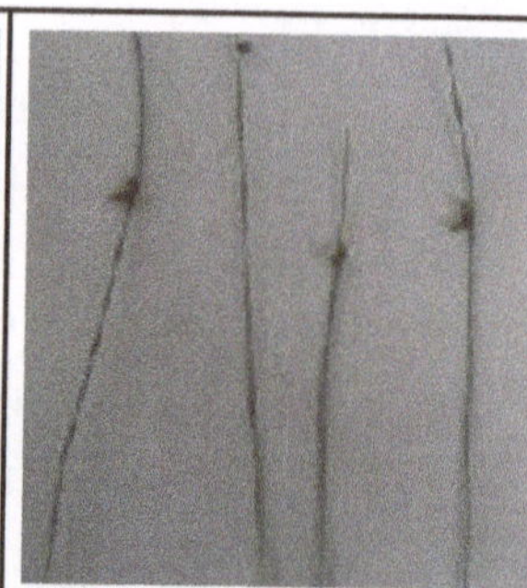
Scirpus

Phalaris Minor

Avena Ludoviciana

Polypogon Monspelensis

Poa Annua

Brija Minor

Lolium Temulentum

Rhaphanus raphanistrum

Chenopodium album

Coronopus didymus

Fumaria Parviflora

Melilotus Alba

Medicago Denticulata

Anagallis arvensis

Ranunculus arvensis

Spergula arvensis

Malva Rotundifolia

Plantago Lanceolata

Stellaria Media

Imperata Cylindrica

Cynodon Dactylon

Pennisetum Polystachion

Saccharum Spontaneum

Fimbristylis Miliaceum

Equisetum Arvense

Ageratum Houstonianum

Oenothera

Erigeron Canedensis

Lantana Camara

Wild Rose

Parthenium Hysterophorus

Bidens Pilosa

Rumex Obtusifolius

Circium Arvense

Chromolaena Adenophorus

Hackalia Uncinata

Ubetia Spps

Euphorbia thymifolia

Alternanthera philoxeroides

Cannabis sativa

Datura Stramonium

Taraxacum Officinale

Urena Lobata

Solanum Nigrum

Ranunculus Sp

Rubus

Xanthium Strumarium

Leucas Aspera

Syndrilla

Conyza Stricta

Polygonum Hydropiper

Solanum Xanthocarpum

Exercise - 3

Morphological Identification of Weeds

Grasses: Leaves usually have a ligule or at times an auricle. Leaves are narrow, arranged in sets of two. The leaf sheaths are split around the stem, with the stem round or flattened in cross section with hollow internodes.

Example : Echinochloa colona, phalaris minor

Sedges: Leaves lack ligule and auricles, and the leaf sheaths are found continuous by around the stem. Leaves are narrow, arranged in sets of three. In many species the stem is triangular in cross section with solid internodes.

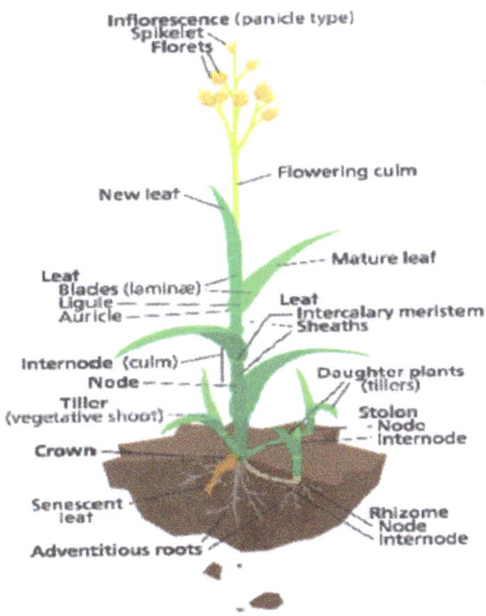

Example : *Cyperus rotundus, Cyperus iria*

Dicots: Plants whose seedlings produce two cotyledons or seed leaves. Dicots are usually typified by netted leaf veination (called reticulate veination) and flowering parts in fours, fives, or multiples thereof. Leaves are wide. They are commonly called broadleaved plants.

Example: Wild mustards, nightshades and morning glory.

Grass Weed Characteristics

Grasses have narrow leaves with veins that run in the same direction and small, not very noticeable flowers. Stems are usually round and have visible bumps or joints where the leaves attach (nodes). Most of the time, they are hollow except at the points.

Growth Habits

Growth habit is a term used to define the kind of shoot growth present in a specific turfgrass. The three fundamental types are stoloniferous, rhizomatous, and bunch-type. Turfgrasses with a bunching habit spread mostly or entirely by tillering. Rhizomatous turfgrasses can spread via below-ground shoots known as rhizomes in addition to tillering, whereas stoloniferous turfgrasses spread by above-ground shoots known as stolons.

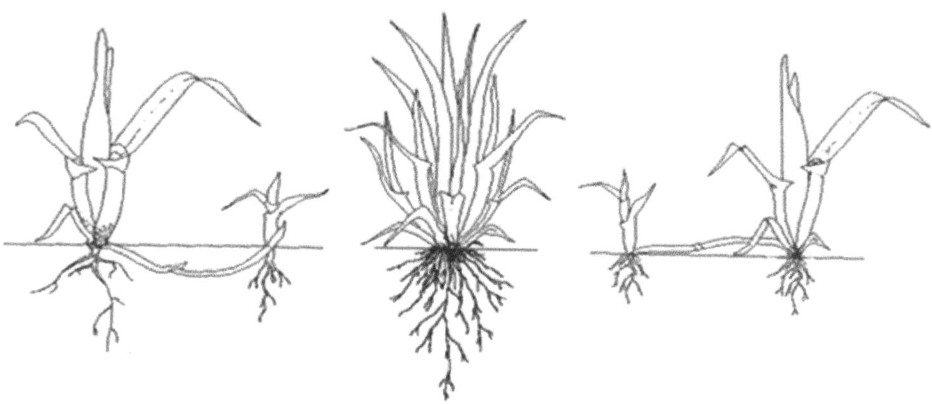

Rhizomatous ***Bunch*** ***Stoloniferous***

Roots

The roots of grass are fibrous. The size of the root system depends on the plant's genes, its environment, and how it is cared for. In general, grasses that can make the most growth at the top can also make the most growth at the roots. Most native grasses have a weight ratio of about 0.8:1 to 1.5:1 between their roots and their tops.

Stems (Culm)

The stem or culm, starts out a single section but develops nodes and internodes that segment it along its length. Stems (culms) grow up from the base of the plant (the crown) and are rounded or angled in cross-section. In most grass species, stems are hollow except where leaves attach to the stem (joints or nodes).

Nodes

Joints along stem where leaves are borne.

Internode

Portion of stem between two successive nodes.

Modified Stems

- **Adventitious stems -** Either rhizomes or stolons.

- **Stolons (runners)** - Stolons are creeping stems that form near the surface of the ground. They produce rooted plantlets at their nodes and ends.
- **Rhizomes** - Rhizomes are below ground creeping stems that often send out roots and shoots from their nodes.

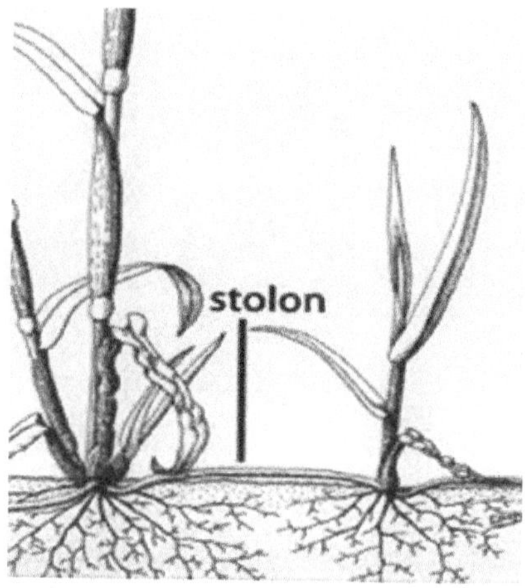

Seedlings (new leaf blades)

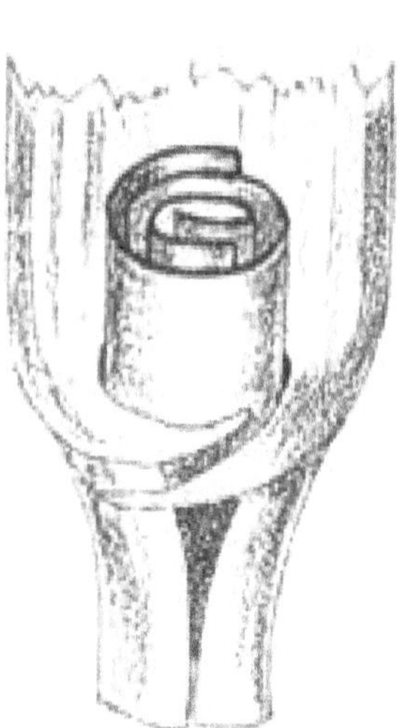

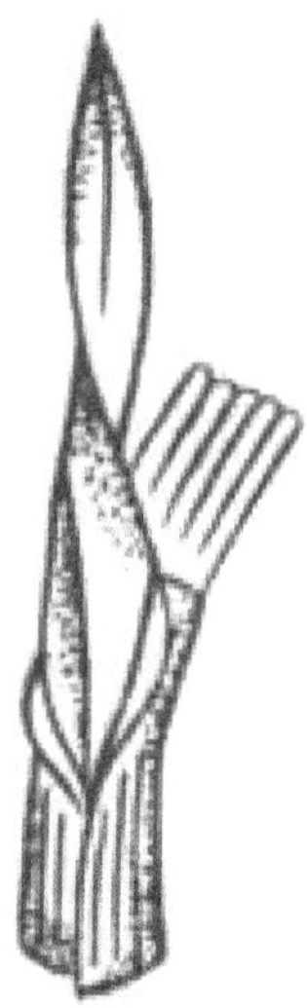

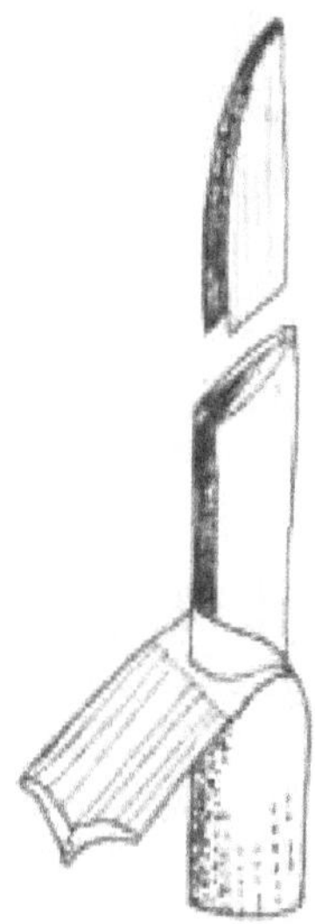

New leaves are either rolled or folded in the bud.

Leaves leaf blades and tips:

Rolled

Leaves may be flat, rolled, or twisted, and vary by width and by the shape of their tips.

Leaf Blade

Leaf blade is a part of leaf the bends away from stem, usually flat to facilitate photosynthesis.

Leaf blades tips

Blunt Shape ***Keeled Shape*** ***Tapering to a Sharp Point***

Collar

The collar of the leaf blade is the ideal spot to look for grasses before they flower (flower heads may provide a more obvious means for identification). On the outside of the leaf, the collar shows where the blade meets the outer sheath. It has the blade, the sheath, the ligule, and the auricles. These parts look different on different kinds of plants, so they can be used to tell them apart.

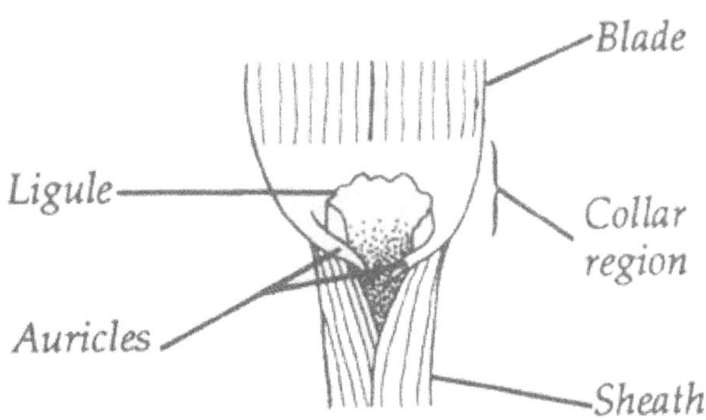

Collar: Region of junction between blade and leaf sheath of grasses.

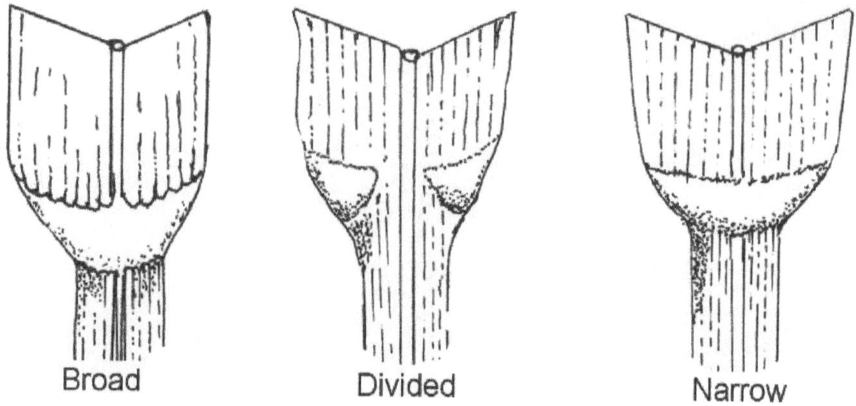

Sheaths

The lower portion of the leaf blade that encircles the stem is the sheath. Sheaths vary in structure and, therefore, are used in identification.

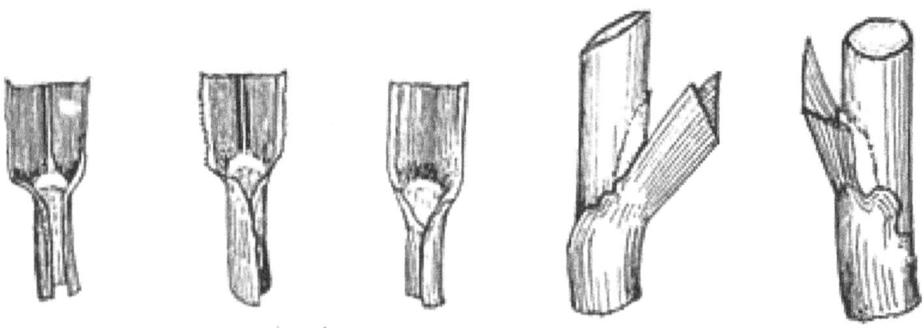

Different Types of Sheath

Ligules

A ligule is an outgrowth from the sheath. The ligule, meaning little tongue, usually clasps the stem firmly on the inside of the leaf at the junction of the sheath and blade, preventing dirt and water from getting between them. In the absence of a seedhead, the ligule is often used to identify a grass. Some ligules are membranous or papery and some are only a ring of hair.

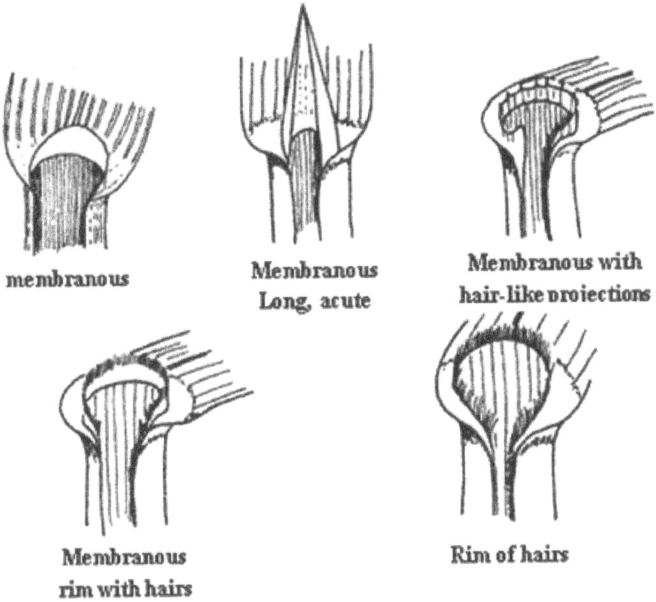

Types of Ligules

Auricles

Auricles are small, clasping outgrowths that may or may not be present on the leaf collar. Because they vary in size and shape, they are used in the identification process. Different types of auricles are given below:

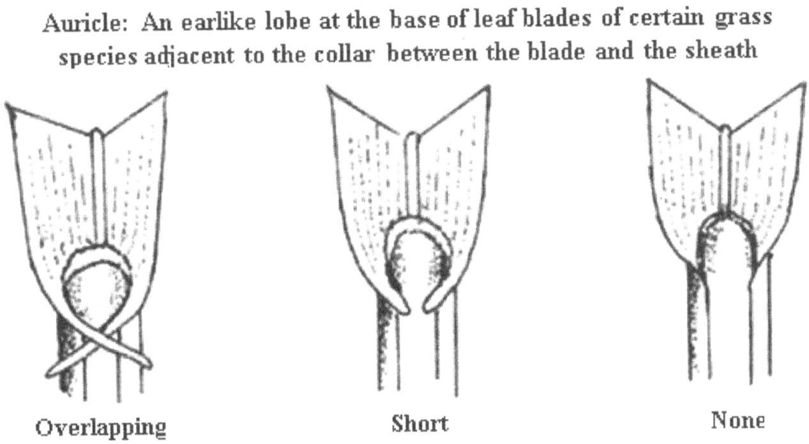

Grass Flower Heads

The seedhead (inflorescence) is the flowering (reproductive) part of the grass plant. Generally, the seedhead has no leaves. On some grasses, a sheathlike bract, called spathe, encloses or partly encloses the seedhead. The spikelet is the basic unit of the seedhead. It may be pediceled (on a pedicel or footstalk) or sessile (without a pedicel). The spikelet consists of a rachilla (jointed stem or axis), one to several florets, and two glumes.

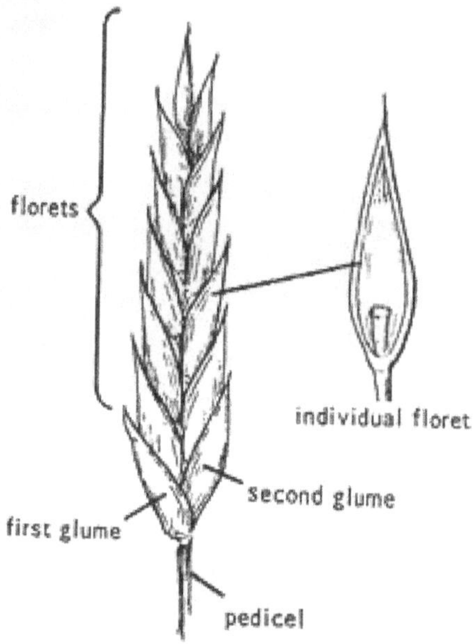

Spikelet having several florets

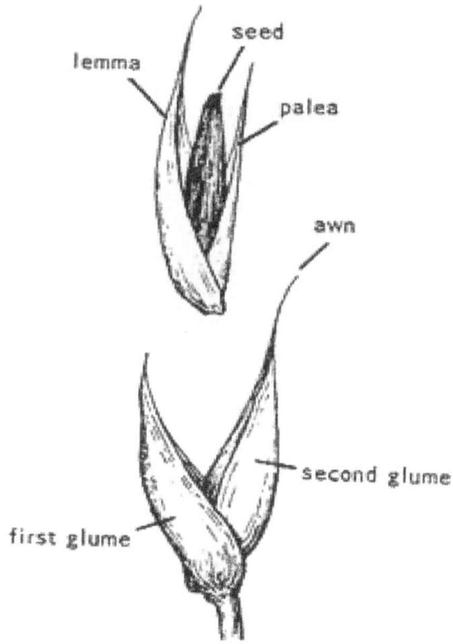

Spikelet having only one floret

The florets are borne in two ranks on the rachilla. Below them are the **glumes** - two bracts without flowers. Each **floret** consists of one flower or seed enclosed in two papery membranes called **lemma** and **palea.**

The lemma is borne on the rachilla above the pair of glumes and the palea at the base of the flower or seed. Both the glumes and lemma have nerves or veins that run from the base to the tip. If the center nerve is extended, it is called an **awn.** Because awns are of different lengths, shapes, and colours, they are often used to identify a particular grass. The basic forms of grass seedheads are **spike, raceme** and **panicle** but they may grade from one form to another and may have specialized forms.

Spike is a seedhead in which one or more sessile spikelets are borne on the main axis **(rachis).**

Raceme is a seedhead in which the spikelets are borne on individual footstalks (pedicels) growing directly on the main axis (rachis).

Panicle is a seedhead with a main axis and subdivided branches. It may be compact and spikelike or open.

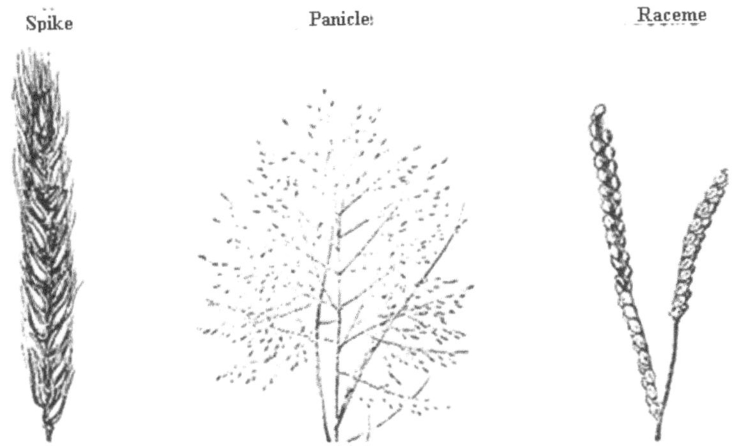

Examples of Spike Flower Heads

	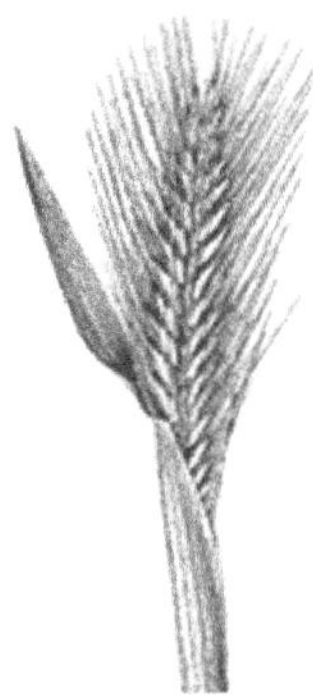	
Examples of Raceme Flower Heads	***Hare Barley***	***Foxtail Barley***

Smooth Crabgrass	***Dallisgrass***	***Fall Panicum***

German Velvetgrass	*Wild Oats*

Sedges Characteristics

Sedges are perennial plants that are commonly found in shallow water or moist soils and can reach 4 feet in height. They resemble grasses and often grow in thick clusters.

Root Systems and Tubers

Sedges have a fibrous root system and may spread by underground rhizomes and/or aboveground stolons. Many sedges have tubers from which new plants can form.

Creeping growth

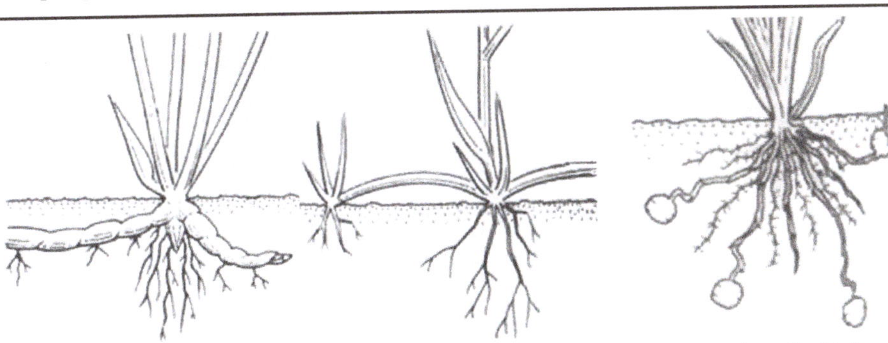

Creeping growth Creeping growth habit by stolons Root system with tubers habit by rhizomes

Sedge stems are usually solid and triangular in cross section, which differentiates them from grasses, which are usually round and hollow (except for where the leaves attach to the stem).

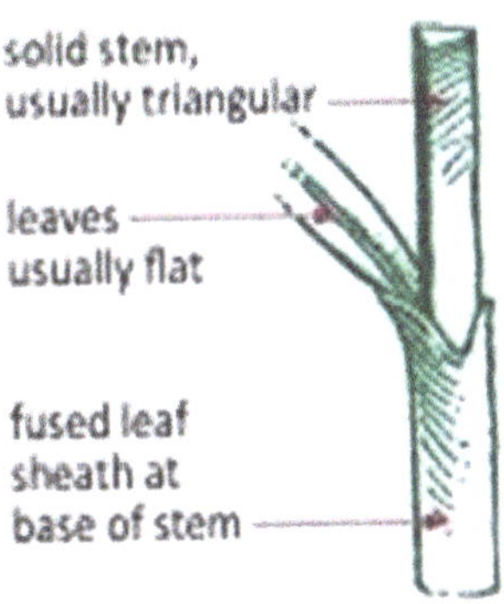

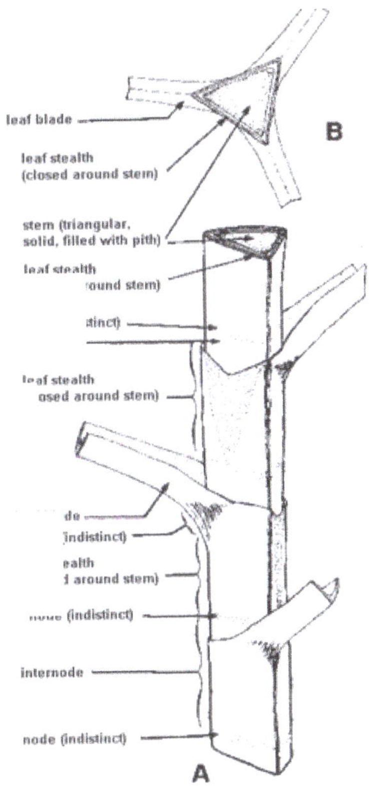

Leaves

Many species have long, thin-textured, narrow, flat leaves which are usually arranged in threes.

Broadleaf Weed Characteristics

Broadleaf plants have relatively broad leaves, whereas leaves of grasses and sedges leaved are bladelike. Leaves of broadleaves have one main vein from which smaller veins branch /come out.

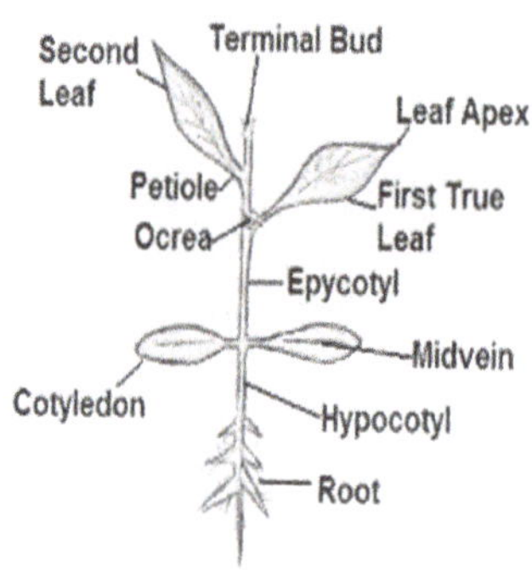

Growth Habit

Broadleaves may grow prostrate and form a mat or they may grow upright.

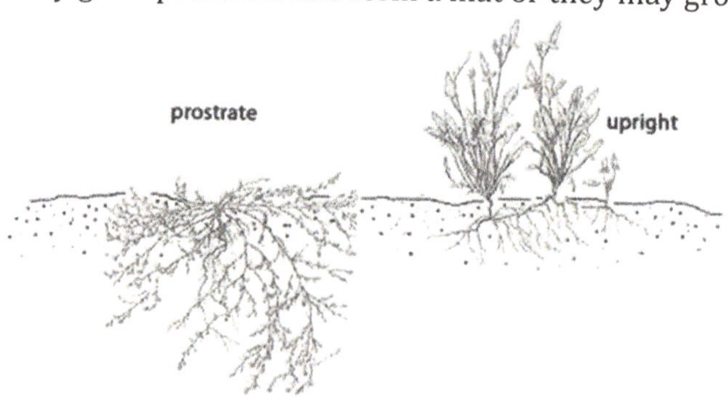

Seedlings

The cotyledon (seed leaf) and first true leaf are major characteristics commonly used in identifying broadleaf weeds.

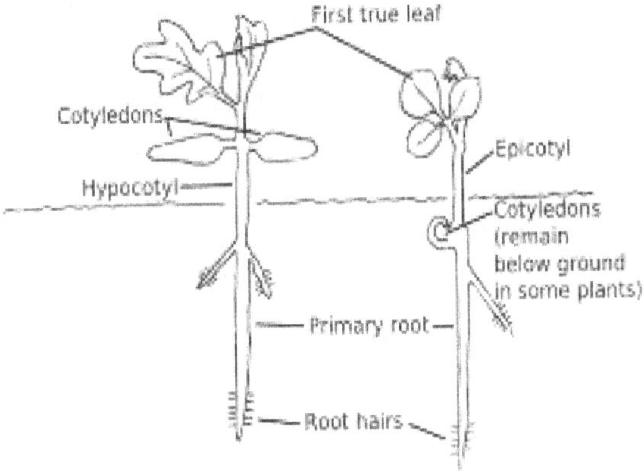

Stems

Stems may be rounded or angled in cross section.

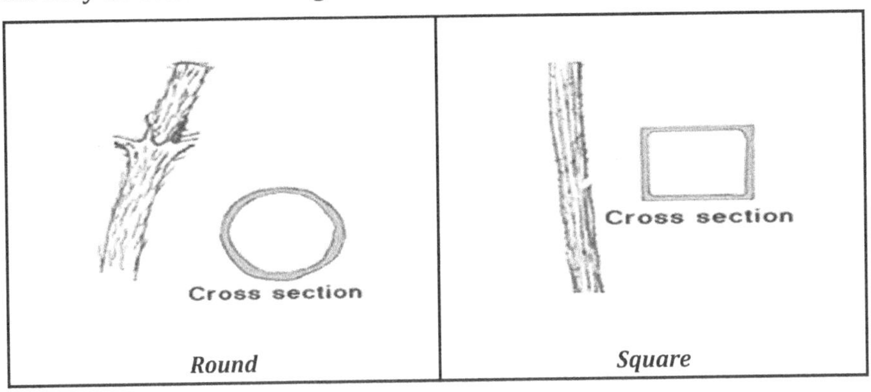

Leaf Arrangements

Leaves are arranged along the stem in a variety of ways according to how they attach at stem joints (nodes).

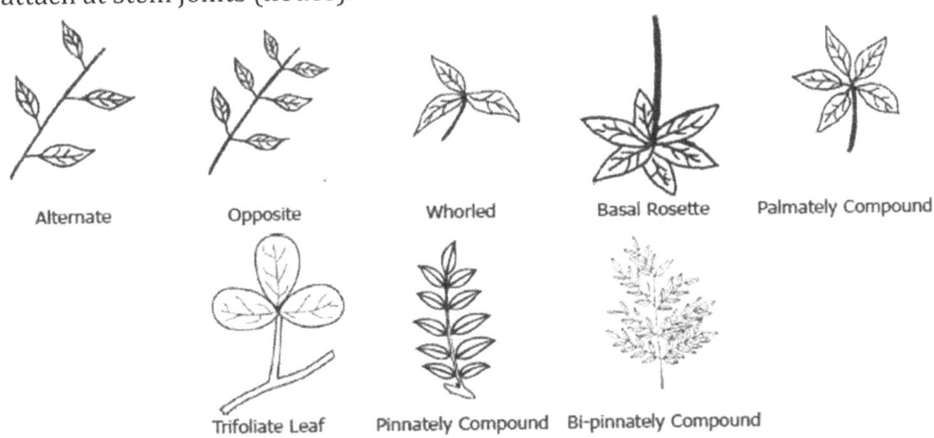

Leaf Margins

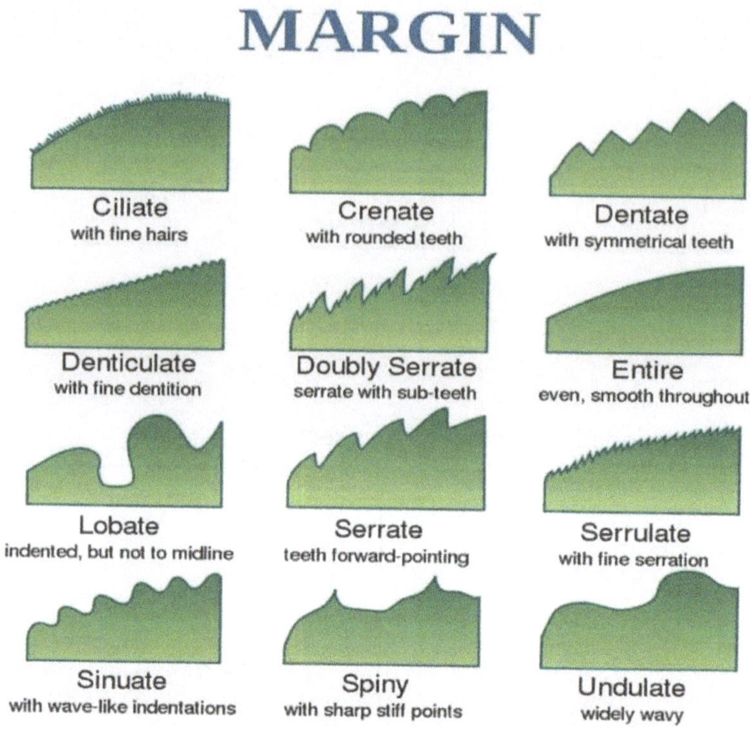

The type of leaf edge (margin) is another way to identify a plant.

Leaf Shapes

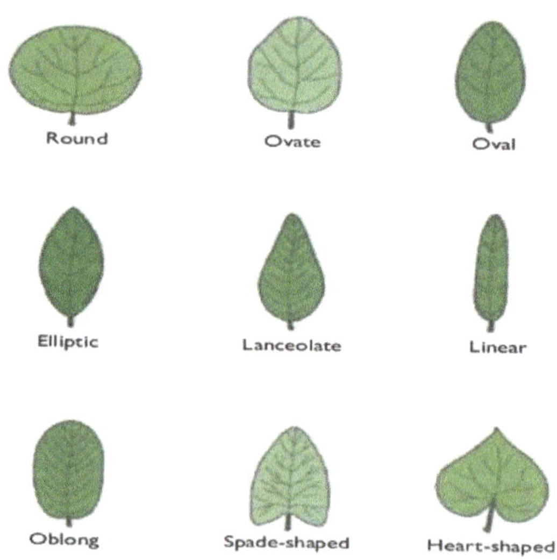

Leaves vary in shape.

Leaf Veins

Depending on the type of plant, leaf veins are either parallel or netted in pattern. In leaves with netted veins, major veins branch from the main ribs and subdivide into finer veinlets. Fruit and deciduous trees, vegetable plants (not corn), most wildflowers, and many shrubs and flowers are examples of plants with netted veins. In leaves with parallel veins, major veins most commonly run parallel to each other along the length of leaf. Less typically parallel veins run laterally from the midrib of the leaf to the leaf edge (not shown).

Examples for parallel veins : Sedges, cattails, lilies, irises, and grasses (e.g., corn, rice, wheat, turf grasses).

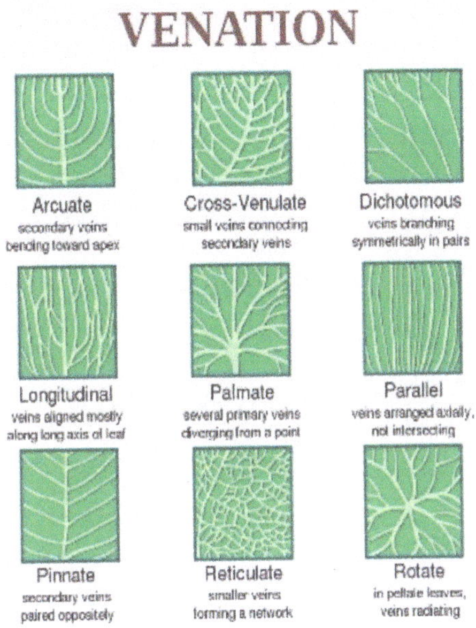

Simple and Compound Leaves

A simple leaf is a leaf blade that is one piece, although it may be deeply lobed, divided or dissected. A compound leaf is a leaf with 2 or more distinct leaflets.

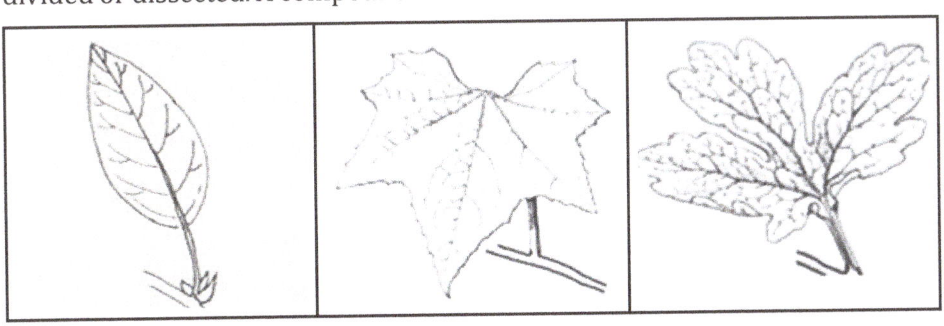

		Simple Leaves
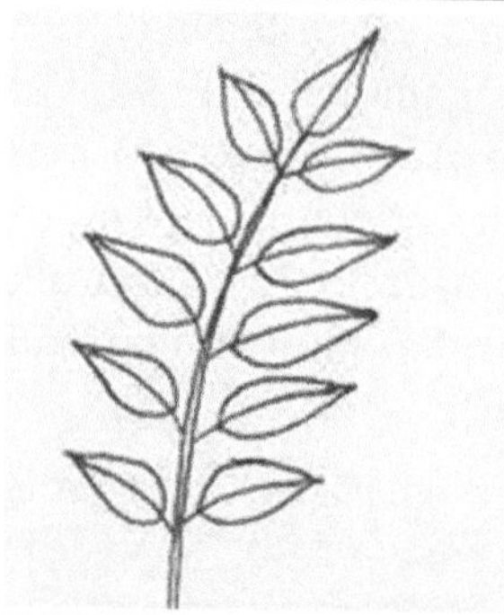		
	Compound Leaves	

Leaf Stalks (Petioles)

Leaves may be attached to the stem by a petiole (short stalk) or may be sessile (attached directly to the stem). Some sessile leaves clasp the stem.

Leaves are attached to the stem by a stalk (petiole)	Leaves are stalkless and attached directly to the stem	Leaves are stalkless, attached directly to the stem, clasping it

Flowers and Flower Heads

Solitary	*Raceme*	*Spike*
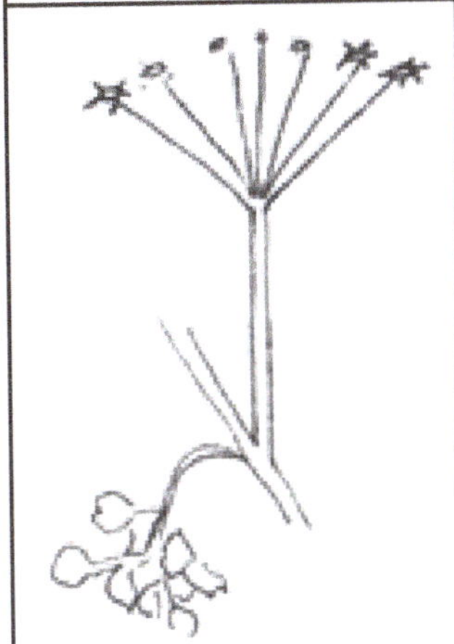	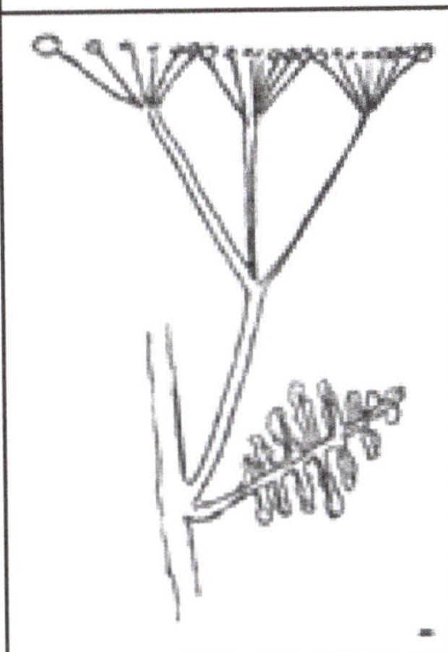	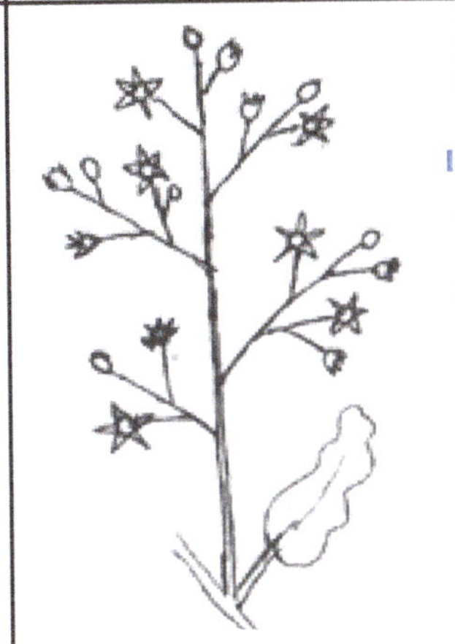
Simple Umbel	*Compound Umble*	*Panicle*
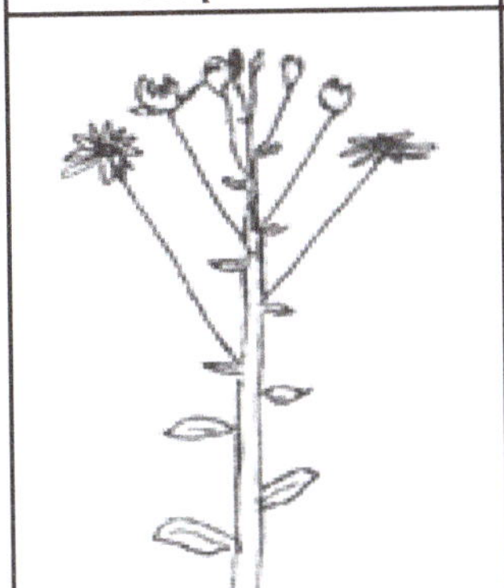		
Corymb	*Head or Capitulum*	

Roots

Broadleaves have a tap root system having a large primary tap root with smaller lateral roots, or a hybrid of both.

Taproot

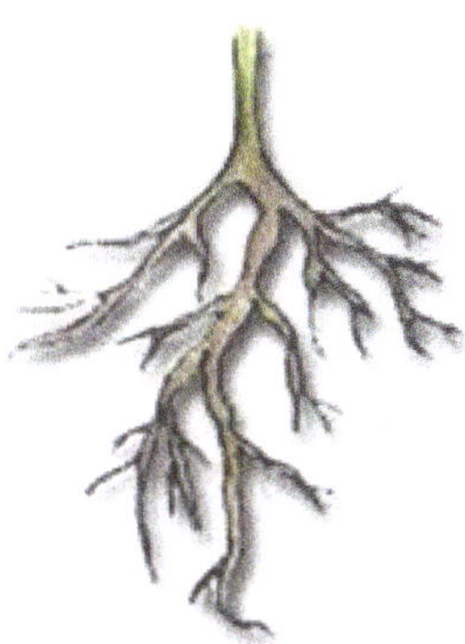

Exercise - 4

Techniques of Weed Preservation

Collection of weeds and their preservation is essential so that one can know about their growth, development, competition and management under field conditions.

Herbarium is a collection of plants that are dried, pressed and preserved on herbarium sheets and arranged in sequence in accordance of specific purpose for future reference, record and study.

Objectives

i. To know about the common weeds,
ii. To learn about collection of weed specimen and
iii. Preparation of weed album/herbarium.

Materials

Herbarium sheets (Appendix), wooden block press, blotting papers or old newspapers, pencil or marker, glue/adhesive tape, thread etc.

Procedure

The scientific method of weed collection and preparation of herbarium consists of the following important steps:

1. Collection of weed sample

The weeds with height of about 15 to 20 cm are uprooted carefully along with roots. The specimen should contain all parts of the weed plant including root, stem, leaves, flowers and fruits. For good identification, no part of the plant should be excluded. If plants are too small or large, extra care may be taken in collecting a specimen. A large plant may be divided into 2 or more sections, each pressed separately. However, excess branches or leaves may be removed provided remaining leaves and branches truly represent the plant. If the plants are very small, more number of specimens of the sample plant should be collected.

2. Pressing and Drying of Collected Specimen

The weed specimens are kept on blotting paper with the leaves, roots, and flowers in the right place and another piece of blotting paper on top. Plants with long stems or leaves can be folded in a V, N, or W shape, but they shouldn't be folded back on themselves. The paper is placed on a flat surface, and enough weight is put on the top blotting paper to press the weeds inside. Overnight, these two blotting papers soak up all of the water from the weed sample. The next day, the weight is taken off, and the weeds are moved to another piece of dry blotting paper. The sample is then turned upside down and covered with dry blotting paper. Again, enough weight is kept so that the sample can be pressed.

Herbarium Press

In this way the paper is changed 2 to 3 times by changing the position, of weed every time to absorb moisture from all parts of weed plant. When the specimens dry completely, it is ready for preservation.

3. Poisoning

The weed specimen once dried need chemical treatment to protect them from insect attack and other destructive organisms. Thus dipping of specimens in an insecticide solution is essential. The following solutions are used for the treatment.

i. Mercuric chloride (25 g) + cresol (25 g) in one litre industrial alcohol.
ii. Mercuric chloride (15 g) + phenol crystals (10 g) in one litre denatured alcohol.
iii. 20% lauryl pentachloro phenate in methylated sprits.

The plants are re-dried as discussed earlier. These solutions are corrosive and proper care is needed to handle them.

4. Mounting on Herbarium Sheet.

After the specimens are dried and poisoned, they are mounted on herbarium sheets. Generally the size of the herbarium sheet is 42 cm x 29 cm (A sample herbarium sheet is given in Appendix section). The herbarium sheet should be of good quality and medium in weight. The specimens are placed in the centre of the sheet. Glue or narrow strips of adhesive tape (preferably transparent) are used to mount the specimens on the herbarium sheet. Several herbarium glues are available in the market. If these are not available prepare the following gum,

i. 500g gum Arabic, 200g sugar and 15g phenol dissolved methylated spirit solution.

ii. Dissolve the 500g gum Arabic in 800 ml boiling water. Add 30g mercuric chloride and 15g phenol. Glue is applied to plant specimen only.

5. Preparation and Fixing of Identification Label

Collection of information

The collector of weed specimen should record maximum useful information at the time of the collection. The data may be written either on the edge of same newspaper in which weed specimen will be brought from field to laboratory or a diary may be used citing some reference number for a particular weed specimen. The collector should record information in respect of the following parameters.

a) **Location:** Name of the village or town nearby the field and its distance and direction from the known town for exact location, the district may be mentioned.

b) **Date:** The date should be clearly mentioned with day, month and year. It should be written as December 27, 2022.

c) **Habitat:** Under this category, name of place with ecological conditions viz., field, pasture, roadside weeds, hillside, sand dune, nallah (eroded stream), light exposure (sun or shade), moisture conditions (dry, moist, wet. etc.) and denseness of community (bare ground, thin or dense population) should be pointed out.

d) **Occurrence of weeds:** A weed under consideration should be described in relativity of number of other species of weeds. For this purpose, an arbitrary scale of comparison using terms like rare, occasional, frequent, common and abundant may be followed.

e) **Noting of essential characteristics nature** - Annual, biennial, perennial Root - Tap, fibrous, adventitious, shallow, deep Stem/branches - Woody, herbaceous, erect, spreading, trailing, prostrate Leaves - Simple, compound, narrow, broad Flower - Shape, colour, fragrance.

f) **Features of special reference:** Some plants in nature are known for their special characteristics in terms of fragrance, colour, leaf curling, stinging hairs, presence of thorns, double colour of leaves, milky juice of stem or leaves, habitat of growth, stickiness etc. The specific characteristics of plant along with right stage of growth and development of the plant should be mentioned.

g) **Miscellaneous points of interest:** A collector by his own wisdom or discussion with local people may collect valuable and rare information about a weed specimen.

LABEL Ref. No:	....
Location and Habitat:	....
Common name (English):	....
(Local):	....
Scientific name:	....
Description:	....
Collectors Address:	....
Date & Time:	....

6. Preservation

The labelled specimens are subsequently organised in weed albums for small collections or in herbarium cabinets for big collections that require long-term preservation. Album is a book with blank pages for storing and organising weed collections. A weed herbarium (plural: herbaria) is a collection of mounted, labelled, and methodically organised preserved weed specimens for scientific research. The phrase can also refer to the structure or room in which specimens are stored, as well as the scientific institution that not only stores but also conducts research with them.

Certain weeds are delicate, heavy, or otherwise incapable of being dried and mounted on sheets. Alternative techniques of preparation and storage may be utilised for these plants. For instance, conifer cones and palm fronds can be preserved in containers with labels. It is possible to pickle representative flowers, fruits, fleshy roots, and stems in formaldehyde to preserve their three-dimensional structure. Typically, weed seeds are air-dried and stored in compact paper/polyethylene envelopes or small glass jars. Regardless of the method of preservation, full details on the location and date of collection, habitat, colour (as it may fade with time), and collector's identity are typically given.

Note: After having complete practical demonstration on weed preservation, each student of weed management will prepare weed album by his own. The students will also update the weed herbarium of the department.

Exercise - 5

Studies on Biology of Important Weeds

The biology of weeds is concerned with their taxonomy, genetics, establishment, growth, and reproduction. Biology is the science of life that involves the study of living organisms, including their chemical compositions, physical compositions, developments, functions and evolution. The word Biology is derived from the Greek word "*bios*" which means 'life' and "*logos*" means 'study'. Modern biology is a vast field which is composed of various branches. Some branches of biology are given in the points below.

- **Botany:** The study of plants and agriculture.
- **Zoology:** The study of animals and their behaviour.
- **Biochemistry:** The study of materials that make up every living thing.
- **Ecology:** The study of the interaction between living organisms and the environment.
- **Genetics:** The study of heredity.
- **Evolutionary Biology:** The study of evolution.
- **Cellular Biology:** The study of cells.

The biology of weeds is mainly about taxonomy, genetics, establishment, growth, and reproduction. Weed biology considers plant attributes such as morphology, seed dormancy and germination, physiology of growth, competitive ability and reproductive biology. Knowledge of weed biology is essential for development of both economically and environmentally acceptable weed management systems. It is also essential to understand and predict how weed species, populations and biotypes evolve in response to the selection pressure primarily due to agricultural and related practices. Weed identification is the first step in understanding their biology.

The students will study biology of 10 most common weeds representing monocots and dicots. They will note down the salient characteristics given below:

(A) Morphology of the plant

i. Leaf: Colour, size, shape, arrangement

ii. Stem: Colour, size, nature (erect, prostrate, angular trailing etc.) nodes and internodes, solid, hollow, woody, tender

iii. Branches: Branched, unbranched, pattern and arrangement

iv. Flowers: Colour, size, type of inflorescence

v. Roots/under ground parts: Shallow, deep, tap root, adventitious, root colour, rhizomes, nuts, bulb etc.

vi. Juncture points: Note the colour, shape, size of any plant part like hair, ligule, auricle, glands etc. present at the joining point of stem with root, leaf with stem, inflorescence with the main stem / branch etc.

vii. Fruits / seeds: Colour, size, shape

viii. Special points: Nature of plant sap (milky, juicy, gum etc.) and its colour, special modification on the plant etc.

(B) Growth duration: Annual / biennial /perennial

(C) Categorization in broad groups: Grasses / sedges / broadleaf

(D) Nature: Associated, crop bound, parasitic, facultative, obligate

Weed Species	Weed Species

1. Echinochloa Colona	*2. Commelina*

3. Phalaris Minor

4. Avena Ludoviciana

5. Anagallis Arvensis

6. Alternanthera Philoxeroides

7. Lantana Camara

8. Parthenium Hysterophorus

9. Cirsium Arvense

10. Bidens Pilosa

Facts

1. *B. pilosa* is used as an anti-diabetic herb in America, Africa, and Asia. R. P. Ubillas, C. D. Mendez, S. D. Jolad et al., "Antihyperglycemic acetylenic glucosides from *Bidens pilosa*," *Planta Medica*, vol. 66, no. 1, pp. 82–83, 2000.
2. Fresh *B. pilosa* is used to treat snake bites and wounds.[S. Dharmananda, "A Popular Remedy Ecapes Notice of Western Practitioners," 2013, http://www.itmonline.org/arts/bidens.htm]
3. Studies conducted in India have found that **Lantana leaves can display antimicrobial, fungicidal and insecticidal properties.** *L. camara* has also been used in traditional herbal medicines for treating a variety of ailments, including cancer, skin itches, leprosy, chicken pox, measles, asthma and ulcers.

Cirsium arvense (Canada thistle) has been used to coagulate milk. Mohegan Indians used it in a mouthwash for infants and in a tuberculosis remedy for adults. Montagnais also use it for tuberculosis. Ojibwa used it as a bowel tonic.

In India, whole plant of *Anagallis arvensis* L. is used as sedative, stimulant and antiasthmatic, and antiflatulent in cattle. [Tiwari, S., Plants: A rich source of herbal medicine. *Journal of natural products,* 2008. 1(0): p. 27-35.]

Exercise - 6

Study on Crop-Weed Competition

Competition is the action of endeavouring to gain what another endeavours the gain at the same time. Weeds compete with crops for nutrients, moisture, light and space. The nature of crop-weed competition is highly crop specific and varies with several weed factors, including the time of weed occurrence, weed density, critical period of weed competition etc. Weeds have an edge over crop plants in respect of depletion of resources. Therefore, the study of crop-weed competition and its effect on crop plants are critical for obtaining higher crop yields. Further, we also need to evaluate or compare the effectiveness of different weed control practices including herbicides using different indices and methods to reach a good decision.

Objectives

i. To estimate the competitive effect of weeds on crop growth.

ii. To show that different crops differ in their ability to compete with each other.

iii. To demonstrate visual effect of competition among weeds and crop plants.

Materials

Large earthen pots or plastic trays, soil, paper, towels, labels, seeds of pearl millet and groundnut, seeds of *Trianthema portulacastrum*, *Amaranthus viridis* etc.

Procedure

After filling up of earthen pots with sand and FYM, place 20 pearl millet seeds and 20 groundnut seeds in separate pots. Put the weed seeds at the rate of 0, 60, 120 and 240 per pot. After emergence, maintain a crop: weed ratio of 1:0, 1:1, 1:2 and 1:4 by thinning. The treatments are to be applied in 3 replications. This experiment may be done in trays as we do in germination tests.

Experiment Details

i. Test crops: Pearl millet and Groundnut
ii. Total treatments: 3 x 4 = 12
iii. Design: C.R.D.
iv. Replications: 3

Observations

Record the data on germination, height, fresh and dry weight of shoots of crop plants and weeds at one week interval upto 6 week of growth as given in Table 17.1.

Result

Analyze the data statistically and prepare a comparative statement that how the competitive ability of both crops differ, what is the effect of increasing density of weeds on crop growth and its canopy development, which crop is having more competitive ability and how do weeds compete amongst themselves etc. Summarize the results as given below:

i. Crop X is more competitive than crop Y because of better germination, plant height and dry matter accumulation.
ii. As the weed intensity increases, there is no effect on dry matter accumulation of crop Y at 3 WAS but crop X shows poor dry matter accumulation up to 5 WAS due to competition with weeds.

Table 6.1 The effect of crop-weed competition on growth of crop and weeds

Treatment	Observation (Mean over 3 replications)							
	Crops plants							
(A) Pearl millet (Crop: weed ratio)	Period of germination (days)	Population (Numbers)		Height (cm)		Fresh weight(g)	Dry weight (g)	Visual observation
1:0 (control)								
1:1								
1:2								
1:4								
(B) Groundnut (Crop weed ratio)								
1:0 (control)								
1:1								
1:2								
1:4								
Treatment	Observation (Mean over 3 replications)							
	Weeds							

(A) Pearl millet (Crop: weed ratio)	Period of germination (days)		Population (Numbers)		Height (cm)	Fresh weight(g)	Dry weight (g)	Visual observation
1:0 (control)								
1:1								
1:2								
1:4								
(B) Groundnut (Crop weed ratio)								
1:0 (control)								
1:1								
1:2								
1:4								

Exercise - 7

Studies on Losses Caused by Weeds

The weeds are harmful in a number of ways. The damage caused by them is summarized point-wise here as below

1. Reduction in farm productivity: Weeds compete with crop plants for water, nutrients & light. They are hardier and vigorous in growth habit due to consumption of large amounts of water and nutrients thus outgrow the crops. They cause huge losses in yield ranging from 30-40% in barley or wheat to even complete crop failure in case of upland rice and vegetables depending upon type and intensity of weed flora and duration of competition. The yield losses due to weeds can be estimated as per the methodology described as below.

Yield Loss (YL)

Grain yield is recorded from the central 25–30 m^2 of the treatment plots in both weedy and herbicide treated reference plots and then yield loss (YL) is calculated as follow:

$$YL = 100\,(1 - Y_{weedy}/Y_{herbicide})(1)$$

where Y_{weedy}, the average yield in un-weeded plots and $Y_{herbicide}$, the corresponding average yield in plots treated with the herbicide.

Yield loss vs. weed biomass

The analysis involves regression of YL on the weed biomass in weedy plots. Use the negative hyperbolic yield loss function proposed by Cousens (1985):

$$YL = ID/(1 + ID/A)(2)$$

where D is the weed biomass, I the initial slope of the curve and A the upper asymptote of the curve.

Note: The students will learn to estimate losses due to weeds at field, regional, state, country and world scale/level

Table 7.1: Estimation of losses due to weeds in major crops

Particular	Rice	Potato
Field level		

Particular	Rice	Potato
District level		
State level		
Country level		
World level		

2. Reduction in quality of produce: When the crop is harvested from a weedy field the seeds of weeds get mixed with the main crop which results in lowering the quality of the produce e.g. seeds of weeds in wheat or gram. Similarly, bundles of leafy vegetables like methi or palak or Dhania or sarson or bharesh contain green plants of weeds. They fetch lower prices in the market. Weeds impart an undesirable flavour to the milk [*Polygonum hydropiper*; Hulhul (*Cleome viscosa* Linn.)], impair quality of wool of sheep [*Hackalia uncinata*, *Bidens pilosa*, *Xanthium strumarium* gokhru (puncture vine, *Tribulus terrestris*)], and cause death of animals due to poisonous nature of seed (*Dhatura stramonianum*). The quality of the livestock products is reduced.

Table 7.2: Weed seed counts in crop seeds (No/kg seed) and the one adhering to livestock and other animals (No/individual)

Crop/Animal	Name of the Weed Species								
	1	2	3	4	5	6	7	8	9
Wheat									
Rice									
Sheep									
Goat									
Cow									
Dog									

Table 7.3: Common Weeds Poisonous to Livestock

Weed	Where generally found	Livestock affected	Symptoms
Horsetail Equisetum arvense	◆ poorly drained soils ◆ low, sandy, acid soils ◆ cultivated fields ◆ roadsides ◆ waste areas ◆ woods	◆ horses ◆ cattle ◆ sheep	◆ symptoms are slow to develop ◆ jaundice ◆ loss of appetite ◆ weakness ◆ staggering gait ◆ excitability ◆ paralysis

Weed	Where generally found generally found	Livestock affected	Symptoms

3.Weeds harbour insect-pests and disease: Weeds either give shelter to various insect pests & disease pathogens or serve as alternate hosts and thus helps in perpetuating the menace from pests and diseases. E.g.: Gall fly of paddy, midge fly of Jowar, leaf minor of soybean and Groundnut, rust of Wheat, tikka of Groundnut, Black rust of wheat, Downey mildew (*Saccharum spontaneum*). Certain weeds cause irritation of skin, allergy and poisoning in human beings.

Table 7.4: Alternate hosts of pest and diseases

Crop	Pest	Alternate host
Red gram	Gram caterpillar	Amaranthus, Datura
Rice	Stem Borer	Echinocholoa, Panicum
Wheat	Black Rust	Agropyron repens
Maize	Downy Mildew	Sacharum spontaneum

4. Secretions are harmful: Heavy growth of certain weeds like quack grass or motha lower the germination and reduce the growth of many crop plants. This is said to be due to the presence of certain phytotoxins in these weeds.

Table 7.5: Make a list of crops/varieties having allelopathic effect on weeds

S.No.	Crop having allelopathic effect	Susceptible weed
1.		
2.		
3.		
4.		
5.		
6.		
7.		

5. Weeds Check the flow of water: Weeds block drainage and check the flow of water in irrigation canals and field channels thereby increasing the seepage losses as well as losses through over flowing, so reduce the irrigation efficiency. The irrigation efficiency is also reduced. With equal flow rate or amount of water at the channel source heads, students will compare four different kinds of irrigation systems having at least 100 m length and note down the flow rate.

Table 7.6: Comparison of flow rate in different irrigation systems

Kind of the channel	Flow rate at the channel head	Flow rate at the field head
Channel		
Channel		
A cemented or polythene lined channel		
Pipe delivery system		

Interpretation

____

____

____

____

____

____

6. Increase in the cost of production: The main objective of tillage is to control weeds. On an average tillage alone accounts for about 30% of the total expenditure. On instances of heavy weed infestations, cost on weeding or chemical control increases. This results in increased cost of cultivation and reduces the margin of net profit. Weeds cause quicker wear and tear of farm implements; they get worn out early and cannot work efficiently unless they are properly sharpened or mended. Being hardy and deep rooted; the tillage implements get worn out early and cannot work efficiently unless they are properly sharpened or mended.

7. Reduce value of the land: Agricultural lands which are heavily infested with perennial weeds like kans (*Saccharum spontaneum*), wild rose (*Rosa acicularis*), wild ber (*Zizyphus jujube*) or other thorny weeds always fetch less price, because such lands cannot be brought under cultivation without incurring heavy expenditure on labour and machinery.

The students will make assessment of biological degraded land within the university or nearby village and adopt adaptive weed management approach to about 400 m^2 area as follow.

1. Establish management goals and objectives for the site;
2. Determine which plant species or populations, if any, block or have potential to block attainment of the management goals and objectives;
3. Determine which methods are available to control the weed(s);
4. Develop and implement a management plan designed to move conditions toward management goals and objectives;
5. Monitor and assess the impacts of management actions in terms of their effectiveness in moving conditions toward these goals and objectives; and
6. Re-evaluate, modify, and start the cycle again.

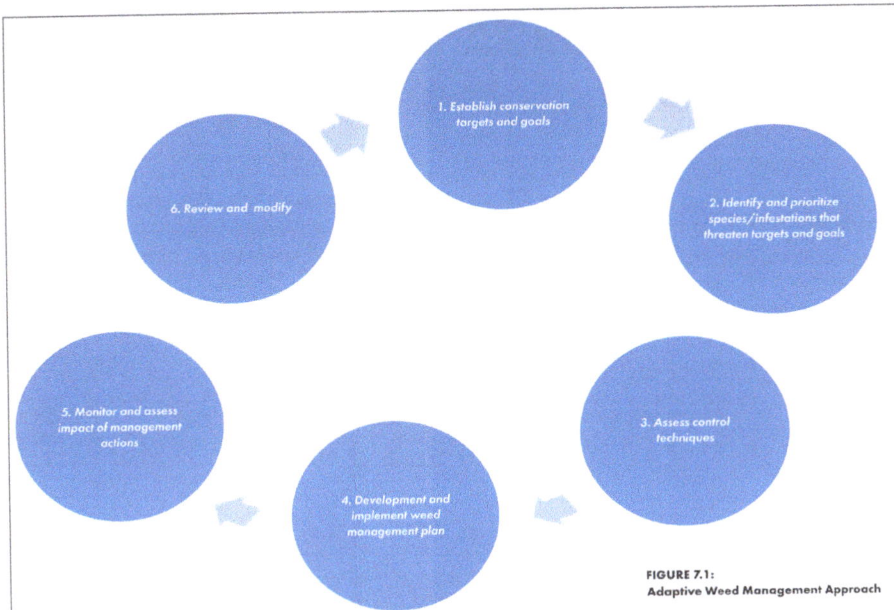

FIGURE 7.1:
Adaptive Weed Management Approach

Note that control activities are not begun until the first three steps have been taken. A weed control program is best viewed as part of an overall restoration program, so focus on what you want in place of the weed, rather than simply eliminating the weed. While selecting control methods, keep in mind preservation of native species, communities and/or functioning ecosystems.

Table 7.7: Step-wise detail of activities

Step	Activity Detail
1.	
2.	
3.	

Step	Activity Detail
4.	
5.	
6.	

Exercise - 8

Study of Weed Population and Quantitative Evaluation of Weed Cover

For accurate estimation of loss caused by weeds and selection of right method of weed control for a given environment, it is essential to evaluate the weed cover in the field.

Objectives

1. To determine the weed population in the field.
2. To estimate the loss caused by weeds.
3. To obtain quantitative data about weed density, frequency, dominance and biomass.

Materials

Quadrat, cloth bags, measuring scale, thread etc.

Procedure

Quadrat sampling is the most common method for obtaining various types of data on weed cover. In fact, quadrat is a sampling unit which has an area of definite size which may be circular, rectangular or square in shape. In general, a square quadrat measuring 1m x 1m is sufficient to represent the composition of an agricultural field. In general, more the sampling units (quadrat), greater is the precision.

Select a field or sample area for study of weed flora. Randomly put the quadrat in different plots and note down the observations as mentioned in Table 18.1.

Table 8.1 Quantitative study of weed flora through quadrat sampling

Plant Species	No. of plants in given quadrat (number)					Total No. of plants	No. of quadrat of occurrence	Total No. of quadrat studied
	1	2	3	4	5			
Cyperus rotundus	10	-	5	-	7	22	3	5
Amaranthus viridis	8	5	3	2	1	19	5	5
Trianthema monogyna	1	2	-	-	10	13	3	5
Grand total	-	-	-	-	-	54	-	-

Formulas for estimation of quantitative parameters.

1. Weed density: Count the number of individuals of particular weed species per unit area and determine the relative density.

$$\text{Density}(\text{D}) = \frac{\text{Total number of weeds in all quadrats}}{\text{Total number of quadrats studied}} \times 100$$

$$\text{Relative Density}(\text{RDe}) = \frac{\text{Number of individuals of a given weed species}}{\text{Total number of individuals of all the weed species}} \times 100$$

2. Weed Frequency: This parameter determines the degree of dispersion of a given weed species in an area.

$$\text{Frequency}(\text{F}) = \frac{\text{Number of quadrats of occurrence of a species}}{\text{Total number of quadrats studied}} \times 100$$

$$\text{Relative Frequency}(\text{RF}) = \frac{\text{Frequency of given weed species}}{\text{Sum of frequency of all weed species}} \times 100$$

3. Weed dominance: It refers to the ground coverage of a given weed species in relations to density of the given weed species.

Dominance (D) = Average basal area of a given weed species x density

$$\text{Relative Dominance}(\text{RDo}) = \frac{\text{Total basal area of a given weed species in all quadrats}}{\text{Total basal area of all weed species in all quadrats}} \times 100$$

OR

$$\text{Relative Dominance}(\text{RDo}) = \frac{\text{Dominance of weed species}}{\text{Dominance of all weed species}} \times 100$$

4. Weed abundance: It can be calculated by the following formula

$$\text{Weed abundance} = \frac{\text{Total number of weed species in all quadrats}}{\text{Total number of quadrats in which the species occurred}}$$

5. Weed cover: Following methods are used to determine ground cover by the weed species:

a) **Braun-Blanquet method:** In this method different scale number are given depending upon the area of coverage by the weed species:

Scale / Class	Description
0	Sparsely present
1	Plentiful but with small coverage
2	Numerous or covering 1/20 of the area
3	Number of weed species covering ¼ to ½ of the area
4	Number of weed species covering ½ to ¾ of the area
5	Weed species covering more than ¾ of the area

b) **Rating Scale:** In this, 0 to 10 point scale is used where 0 means no cover and 10 = 100% ground cover.

6. Weed biomass: Relative dry weight (RDW) of weeds can be calculated by the following formula:

$$\text{Relative dry weight (RDW)} = \frac{\text{Dry weight of given weed species}}{\text{Total dry weight of all weed species}} \text{X100}$$

7. Importance Value Index (IVI): Importance Value Index is used to express the overall dominance and ecological success of a given weed species over others in a community with a single value.

$$\text{IVI} = \text{Relative density} + \text{Relative frequency} + \text{Relative Dominance}$$

Result

Record the data for individual weed species in a tabular form. Compare, correlate and interpret the competitive effects of weeds through indices and formula.

Exercise - 9

Studies on Indigenous Technological Knowledge (ITKS) on Weed Management

What is Indigenous Knowledge?

Indigenous knowledge (IK) generally refer to knowledge systems embedded in the cultural traditions of regional, indigenous, or local communities. IK is the knowledge that an indigenous (local) community accumulates over generations of living in a particular environment. This definition encompasses all forms of knowledge – technologies, know-how skills, practices and beliefs – that enable the community to achieve sustainable livelihoods. A number of terms are used interchangeably including indigenous knowledge (IK), Traditional Knowledge (TK), traditional ecologic knowledge (TEK), Indigenous Technical Knowledge (ITK), Local Knowledge (LK) and Indigenous Knowledge System (IKS). Traditional knowledge includes types of knowledge about traditional technologies of subsistence (e.g. tools and techniques for hunting or agriculture), midwifery, ethno botany and ecological knowledge, celestial navigation, ethno astronomy, the climate etc. These kinds of knowledge are crucial for the subsistence and survival and are generally based on accumulations of empirical observation and interaction with the environment.

Halod an indigenous method of weed control in direct seeded rice

Some of the ITKs in relevance to weed control as reported in the literature are summarized as below

1. In weed management farmers knew of hand weeding/ picking and the use of ash to control weeds/ pests. Ash is spread/ broadcasted over field for weed control, surface soil crust breaking, and making soil more porous by conserving the rainwater.
2. To save ratoon crop of sugarcane, farmers use sugarcane leaves for mulching in between the rows. After first rain in June, they apply 50 kg urea per acre as top dressing for fast rotting of the sugarcane leaves. By this process farmers save sugarcane crop from weeds and minimize the cost on inter culture operations.
3. To minimize the weeds and as well as increase nitrogen level in field, farmers grow sesamum or sesbania crop.
4. Farmers grow sunhemp in those fields where the motha weed is problem.
5. Summer ploughing was practiced to get rid of weeds, killing soil borne insects and pathogens, to eliminate/eradicate non-crop plants in the field and to enrich soil nutrients potassium. The fact is that the soil burning was a practice to clean field in short time. About 79.17 per cent of the farmers practiced summer ploughing in the summer fallows. Farmers believed that resting stages of insects and pathogens get destroyed when exposed to sunlight, weed growth could be eliminated by deep and frequent summer ploughing and hardening of laterite soil with the result of moisture reduction, deep ploughing loosen the soil and help in soil moisture conservation.

6. Summer ploughing is done for weed control & also for controlling insects-pests.
7. Hand weeding is used for weed control.
8. For destruction of all type weed seeds, before sowing the crop irrigation & ploughing method is used.
9. Burning of weeds is also a common practice of weed control.
10. Common salt is used for Lucerne seed treatment to control "doddar" (Cuscuta reflexa), also known as "Amar Bel."
11. The ash of cow dung is dusted on the plants at the rate of 50-60 kg/ha to prevent growth of cuscuta.
12. "Weed Collector", made of wood and iron is used for weed collection from the field.
13. Traditional "Indigenous/desi plough" is a most common multipurpose implement.
14. "Blade Harrows" is a useful implement for weeding and inter-cultivation in the standing crop.
15. Weeding is done by "Kudaali/Spade or Khurpi."(Small hoe with wooden handle)
16. Most farmers are aware of hand weeding/picking (64.7%) and use of ash (69.3%) the ways of controlling weeds/pests on the Farms.
17. The ITK that was least known to farmers was the use of goat waste (12.0%) as a weed/pest control agent. The number of goats in the study area has significantly reduced, as farmers prefer to keep cattle instead of goats.

Some indigenous technological knowledge from Tamil Nadu [http://agritech.tnau.ac.in/itk/IndigenousTechKnowledge_Oper.html#WEED%20MAN AGEMENT]

- If weeding is not done, it will reduce three fourths of crop yield.
- Weeding is not required under dry land condition. If weeding is not done the weed growth is controlled naturally and it also helps to conserve moisture.
- Repeated ploughing will reduce weed population.
- Crop yield will be less in the fields having 'Arugu' (*Cynodon dactylon*) weeds.
- To control "Arugu1 (*Cynodon dactylon*) grass in ' black soils the field is kept fallow for 3 years.
- Cultivating rice once in three years in garden lands to control 'Arugu' (*Cynodon dactylon*) weeds.

- 'Raising and ploughing the green manure crops tike dhaincha, (*Sesbania* sp.), kolingi (Tephrosia purpurea) in the field before their flowering to reduce weed population.
- Raising *Calotropis gigantea* as a green manure to check the growth of Aarai (Mars/teaquadrifolia) weed.
- Growing horse gram to control nut grass (*Cyperus rotundus*).
- Growing cowpea as a green manure to control nut grass.
- Allowing swine in the fields to eradicate nut grass.
- Frequently ploughing the fields by wooden plough made up of neem trees and frequent application of neem cake in the soil to control nut grass.
- Dissolving 1 kg. of salt and 100g. of sarvodaya soap in 10L of water and spraying this solution to control all the weeds except nut grass.
- To control nut grass in the field 50kg. of neem cake is applied both at the time of ploughing and sowing.
- Dissolving 200 g. of salt in 1 L of water and spraying to eradicate congress weed (Parthenium hysterophorus).
- Continuous submergence of field for some time controls the weeds.
- Keeping the irrigation channels free from weeds,

Based on the ITKs mentioned as above, the students in consultation with the instructor concerned will design a survey schedule keeping into consideration the ITKs related to preventive, cultural, physical and mechanical and cropping and competitive methods all included in the schedule. After pretesting the schedule, each student will survey at least two farmers in a nearby village. The compilation and interpretation of the data will be done in the class along with the teacher concerned.

Table 9.1: ITKs used by the respondents (No and %) in different crops

ITK Related to	Crop-1			Crop-2			Crop-3		
	M	S	Me	M	S	Me	M	S	Me
Preventive Weed Control									

ITK Related to	Crop-1			Crop-2			Crop-3		
	M	S	Me	M	S	Me	M	S	Me
Cultural Weed Control									
Physical and Mechanical									
Cropping and Competitive									

M, marginal; S, small; Me, medium; Results and interpretation

Exercise - 10

Weed Control Efficiency and Weed Index

Weed control efficiency (WCE), weed control index (WCI) and weed index provide a logistic support in impact assessment, interpretations and drawing appropriate conclusions in weed management research.

Weed control index is calculated by replacing weed count with weed dry weight.

$$\text{WCE} = \frac{\text{Weed count in control (unweeded) - Weed count in a treatment}}{\text{Weed count in control (unweeded)}}$$

Weed control index is calculated by replacing weed count with weed dry weight.

$$\text{Weed index} = \frac{\text{Yield from weed free - Yield of particular treatment}}{\text{Yield of weed free}} \times 100$$

Weed index is the measure of the efficiency of a particular treatment when compared with a weed free treatment. It is expressed as percentage of yield potential under weed free. More conveniently weed index is the percent yield loss caused due to weeds as compared to weed free check. Higher weed index mean greater loss.

EXERCISE

Example 1. Calculate weed index (WI), and weed control index (WCI) of all treatments, from the data given in Table below.

Treatment	Weed dry weight (g/m^2)	Grain yield (t/ha)
T_1- Bispyribac 20 g/ha	50.9	2.8
T_2- Bispyribac 30 g/ha	46.9	2.93
T_3- Bispyribac 40 g/ha	33.4	3.37
T_4-Weed free	27.5	3.56
T_5- Butachlor 1.5 kg/ha	37.8	2.86
T_6- Pretilachlor 0.6 kg/ha	56.9	2.55
T_7- Weedy check	100.6	0.98

Solution

a. Calculation of Weed Index(%)

i. $\text{WI of } \frac{T_1 = 3.56 - 2.8 \text{x} 100}{3.56} = 21.3$

ii. $\text{WI of } \frac{T_2 = 3.56 - 2.93 \text{x} 100}{3.56} = 17.7$

iii. $\text{WI of } \frac{T_3 = 3.56 - 3.37 \text{x} 100}{3.56} = 5.3$

iv. $\text{WI of } \frac{T_5 = 3.56 - 2.86 \text{x} 100}{3.56} = 19.7$

v. $\text{WI of } \frac{T_6 = 3.56 - 2.55 \text{x} 100}{3.56} = 28.6$

vi. $\text{WI of } \frac{T_7 = 3.56 - 0.98 \text{x} 100}{3.56} = 72.4$

b. Calculation of Weed Control index(%)

i. $\text{WCI of } \frac{T_1 = 100.6 - 50.9 \text{x} 100}{100.6} = 49.4$

ii. $\text{WCI of } \frac{T_2 = 100.6 - 46.9 \text{x} 100}{100.6} = 53.4$

iii. $\text{WCI of } \frac{T_3 = 100.6 - 33.4 \text{x} 100}{100.6} = 66.8$

iv. $$\text{WCI of } \frac{T_4 = 100.6 - 27.5 \text{x} 100}{100.6} = 72.7$$

v. $$\text{WCI of } T_5 = \frac{100.6 - 37.8}{100.6} \times 100 = 62.4$$

vi. $$\text{WCI of } \frac{T_6 = 100. - 56.9 \text{x} 100}{100.6} = 43.4$$

Exercise - 11

Calculation of Weed Infestation, Weed Index, Weed Control Efficiency and Weed Smothering Efficiency

Various indices are used to study the effect of weed density, growth and suppression effect of weeds on crop plants. The common indices used for weed flora study are weed infestation, weed index, weed control efficiency and weed smothering efficiency.

Objectives

1. To study the weed and crop responses to weed control treatments.
2. To study the efficacy of herbicide.
3. To determine the efficacy of intercropping systems in suppression of weeds.

Materials

Experiment plots, weighing balance, oven, weed research papers etc.

Procedure

Record the data on dry-matter production of weeds in unweeded and treated' plots or from intercropped plots and sole crop plots, crop yield from unweeded and treated plots of the experimental area and thereafter compare the data on various indices.

1. Weed infestation: It refers to the percentage of weeds in the composite population of weed and crop plants.

$$\text{Weed infestation}(\%) = \frac{\text{Total number of weeds in an unit area}}{\text{Total number of weed and crop plants in the same area}} \times 100$$

C

2. Weed Index: It refers to the reduction in crop yield due to the presence of weeds in comparison to weed-free plots.

$$\text{Weed index} = \frac{\text{Yield from weed free (hand weeded) plots - yield from treated plots}}{\text{Yield from weed free (hand weeded) plots}} \times 100$$

This is used to assess the efficacy of a herbicide. Lesser the weed index, better is the efficiency of a herbicide.

Example: Calculate the weed index for atrazine and simazine applied in the maize from the following data :

1.Yield of maize from weed free plot	=	20 q/ha
2.Yield of maize from atrazine treated plot	=	18 q/ha
3.Yield of maize from simazine treated plot	=	19 q/ha

Calculation

i. $\text{Weed index for atrazine} = \frac{20-18}{20} \times 100 = 10\%$

ii. $\text{Weed index for simazine} = \frac{20-19}{20} \times 100 = 5\%$

Result

The weed index is less (5%) for simazine compared to atrazine (10%).Therefore, simazine to be applied in maize for crubing the weed biomass production.

3. Weed Control Efficiency (WCE): It indicates the percentage reduction in weed population or dry weight of weeds under treated plot (herbicide) in comparison to untreated plot (weedy). This index is used to compare the different weed control treatment. It is also known as Weed Control Index (WCI). Higher the WCE, better is the herbicidal / weed control treatment.

$$\text{WCE}(\%) = \frac{\text{Weed count in unweeded plots} - \text{Weed count in treated plots}}{\text{Weed count in unweeded plots}} \times 100$$

OR

$$\text{WCE}(\%) = \frac{\text{Dry matter of weeds in unweeded plots - Dry matter of weeds in treated plots}}{\text{Dry matter of weeds in unweeded plots}} \times 100$$

Example: In a weed control experiment in wheat crop, dry-weight of weeds in unweeded plots was 500 kg/ha, whereas in isoproturon and metsulfuron treated plots, it was 250 and 150 kg/ha, respectively. Find out the WCE and report which one herbicide is better.

Solution:

i. $\text{WCE for isoproturon} = \frac{500-250}{500} \times 100 = 50\%$

ii. $\text{WCE for metsulfuron} = \frac{500-150}{500} \times 100 = 70\%$

Result

Since the WCE of metsulfuron is higher (70%) than of isoproturon (50%), so metsulfuron is better than isoporoturon in controlling weeds in wheat crop.

4. Weed Smothering Efficiency (WSE): This index is used to determine the effect of intercropping on suppression of weeds in comparison to sole crop stand.

$$\text{WSE}(\%) = \frac{\text{Weed dry weight in sole crop stand plots - Weed dry weight in intercropped plots}}{\text{Dry matter of weeds in sole crop stand Plots}} \times 100\text{a}$$

Example: Calculate the WI and WCE from the following data in wheat crop

Treatment	Grain yield (kg/ha)	Dry weight of weeds (g/0.25 m2) 20 DAS*				Weed index (%)	WCE (%) 20 DAS
		Sedges	Grasses	Broad leave	Total		
Pendimethalin 1.0 kg/ha	4629	0.25	0.25	0.15	0.65	13.6	70.5
Isoproturon 0.75 kg/ha	4957	0.15	0.15	0.10	0.40	7.5	61.8
Fenoxaprop-P-ethyl 0.075 kg/ha	5343	0.18	0.18	0.15	0.48	0.2	78.2
Sulfosulfuron 0.05 kg/ha	5271	0.18	0.20	0.10	0.48	1.6	76.2
Weed free	5357	0.15	0.20	0.15	0.50	0.0	77.3
Unweeded control	2929	1.0	0.75	0.45	2.20	45.3	-

* DAS=Days after sowing

It is obvious from the data that there was 45.3% reduction in yield of wheat, when weeds were not controlled throughout the season. But the reduction in wheat yield was minimum (0.2 and 1.6%) when fenoxaprop-p-ethyl (0.075kg/ha) and sulfosulfuron (0.05kg/h) were applied as post emergence being comparable to weed free treatment due to effective weed control under both the former herbicides i.e. 78.2 and 76.2% respectively.

Exercise - 12

Studies on Shift of Weed Flora in Long Term-Experiments

A weed shift is the change in the composition or relative frequencies of weeds in a weed population (all individuals of a single species in a defined area) or community (all plant populations in a defined area) in response to natural or human-made environmental changes in an agricultural system.

- Weed shifts occur when weed management practices do not control an entire weed community or population.
- Some species or biotypes are killed by (or susceptible to) the weed management practice, others are not affected by the management practice (tolerant or resistant), and still others do not encounter the management practice (dormant at application).

Examples of weed shifts that have occurred in recent history

Community Shift in Response to Herbicide Use

- Under mid hill conditions of Himachal Pradesh, with use of clodinafop in wheat weed flora was mainly composed of *Poa annua*.
- Similarly in maize, continuous use of pre-emergence atrazine gives subsequent flushes of
- *Commelina banghalensis*, *Brachiaria ramosa* and *Ageratum conyzoides*.

Community Shift in Response to Tillage Change

- Change from conventional to reduced tillage system often causes weed community shifts that include increases in summer annual grasses and small-seeded summer annual broadleaves, winter annual, biennial, and perennial species, and decreases in large-seeded summer annual species.

Community shift as a result of new localized or long-distance introductions

- Under Indian conditions, *Phalaris minor* in wheat and invasion of grasslands and pastures and other non-cropped areas with *Lantana* and *Parthenium* are typical examples.

Practices Influencing Weed Shifts

- There are many factors that interact and influence the weed population dynamics in cropping system.
- Changes in tillage practices, cultural practices, row spacing and planting date, and weed management practices have all had an impact on the weed spectrum.
- Further, the ever changing weather patterns only complicate matters.

Lesson to Learn

- Such shifts are likely to occur in other production systems that suggest that changes in weed flora need to be monitored continuously in all cropping systems and agro-ecological regions in order to assess emerging weed problems and plan weed management strategies accordingly.

Simple proportional occurrence can be used to interpret the results of the initial, middle and current invasion in the particular season (*kharif or rabi*) as follow:

Season	Treatment	Weed flora			Relative proportion (%)		
		Initial years	Middle years	Current survey	Initial	Middle	Current
Kharif							

Rabi							

Note: Use extra sheet

Interpretation

____

____

____

____

____

____

Exercise - 13

Studies on Herbicide Formulations and Mixtures of Herbicides

Chemicals used for weed control are called as weedicide or herbicide. Herbicide formulation refers to the conversion of technical ingredient into commercial product by combining or mixing with the liquid or solid carrier so that it can be applied uniformly. Formulation affects the solubility, volatility, specific gravity and eventually phytotoxicity of the herbicide.

Objectives:

- To study the different types of herbicides
- To know about the trade names and various formulation of herbicides
- To get acquainted with the sources and uses of different herbicide formulations.

Types of Formulations

Most herbicides molecules are directly not in a form suitable for their field application for weed control. These must be mixed with suitable adjuvants to improve their solubility in water, their application feasibility, their leaf surface activity, storage life and safety to users. This is called then a formulation. The different types of formulations are given below:

1. Emulsifiable Concentrate (E.C.):

The herbicide is a product of the active ingredient (Technical material), a solvent and an emulsifier. As active ingredient is insoluble in water, it is prepared by dissolving the a.i. in a small quantity of a solvent and emulsifying it with water to the required degree of concentration eg. Basalin 45 EC.

2. Wettable Powder (W.P.)

Herbicide materials of low solubility may be milled into fine powder that makes stable suspension in water. Wettable powders require continuous agitation to prevent their setting and to give a uniform level of chemical in the spray e.g. Atrazine 50 % WP.

3. Soluble Powder (S.P.)

These can be dissolved in convent amounts of water and sprayed efficiently. Salts of most herbicides are soluble in water eg. Dalapon and sodium salt of 2,4-D etc.

4. Soluble Concentrate (S.C.)

Herbicides which are available in the form of soluble liquids and can be easily added to water, e.g. Dicamba and 2,4-D .

5. Granules (G.)

The granules are small pellets formed from various inert clays and sprayed with a solution of the toxicant to give the desired content. After the solvent has evaporated the granules are packed for use, eg. Butachlor granules.

Table 13.1: Trade name, formulation, a.i. and common use and source of herbicides

S. N.	Common name	Trade name	Formulation and a.i.	Common use	Source
Herbicides					
1.	2,4-D (amine)				
2.	2,4-D (ester)				
3.	2,4-D (Na salt)				
4.	Acetachlor				
5.	Anilofos				
6.	Alachlor				
7.	Atrazine				
8.	Benthiocarb				
9.	Butachlor				
10.	Chlorimuron				
11.	Chlorosulfuron				
12.	Dalapon				

S. N.	Common name	Trade name	Formulation and a.i.	Common use	Source
13.	Diclofop-methyl				
14.	Diuron				
15.	Fluchloralin				
16.	Fluazifop				
17.	Glyphosate				
18.	Imazethapyr				
19.	Isoproturon				
20.	Mestsulfuronmethyl				
21.	Metolachlor				
22.	Metribuzin				
23.	Oxadiargyl				
24.	Oxadiazon				
25.	Oxyflourfen				
26.	Paraquat				
27.	Pendimethalin				
28.	Pretilachlor				
29.	Sulfosulfuron				
30.	Trifulralin				
Formulated Combinations					
1.	Bensulfuron + pretilachlor				
2.	Clodinafop + metsulfuron				
3.	Imezethapyr + imazamox				
4.	Metsulfuron methyl + chlorimuron ethyl				
5.	Mesosulfuron + idosulfuron				
6.	Pendimethalin + imazethapyr				
7.	Sulfosulfuron + metsulfuron- methyl				

Exercise - 14

Studies on Herbicide and Nutrient Compatibility

Compatibility is the capacity for two systems/things to work/mix together without having to be altered to do so. Compatibility of herbicides with fertilizers is tested with 'Jar Test'. Herbicides and fertilizers are normally tank mixed to save time and application cost. However, some herbicides may not be compatible with fertilizers or may be affected by the spraying water.

The product label may state if it can be mixed with a fertilizer or not. If the label does not state any recommendation, one must perform his own compatibility test (Jar Test) before large batches are mixed.

Incompatibility can be both chemical and physical. Chemical incompatibility may result in change of herbicide properties, and this may not be very obvious. Physical incompatibility, on the other hand, can cause the formation of lumps, precipitates, gels or foams. The result may be total loss of the products and use of the tank and/or sprayer. Incompatibility also can reduce effectiveness or cause crop injury. Surfactants and compatibility agents may be added to help maintain component dispersion and improve mixing.

The water source is very important because it may be more difficult to mix with water that is hard and contains large amounts of minerals compared to water with fewer minerals. Fertilizers can influence the pH of the mixture (some raise the pH, and some lower the pH) and thus the suitability of a herbicide with a fertilizer. Avoid letting herbicide/fertilizer mixtures sit overnight because the pH of the mixture can change and cause precipitation.

Because of the above mentioned issue, it is always prudent to conduct a "Jar Test" before mixing any herbicide with a fertilizer to ensure compatibility. This can save time, money and the spray equipment. The following "Jar Test" should be used to test the physical compatibility of herbicides with fertilizers.

Simultaneous or sequential application of herbicides, insecticides, fungicides, antidotes, fertilizers etc., is followed in a single cropping season. These chemicals may undergo a change in physical and chemical characters, which could lead to enhancement or reduction in the efficacy of one or more compounds. The interaction effects were seen much later in the growing season or in the next season due to build-up of persistent chemicals or their residues in the soil. Knowledge on the interactions of various chemicals can be helpful in the formulation and adoption of a sound and effective plant protection programme. It can also help to exploit the synergistic and antagonistic interactions between various pesticides for an effective eradication of weed and other pest problems. When two or more chemicals accumulate in the plant, they may interact and bring out responses. These responses are classified as additive, synergistic, antagonistic, independent and enhancement effects.

i. **Additive effect:** It is the total effect of a combination, which is equal to the sum of the effects of the components taken independently.
ii. **Synergistic effect:** The total effect of a combination is greater or more prolonged than the sum of the effects of the two taken independently e.g. The mixture of 2,4-D and Chlorpropham is synergistic on monocot species generally resistant to 2,4-D. Similarly, low rates of 2,4-D and Picloram have synergistic response on *Convolvulus arvensis*. Atrazine and Alachlor combination, which shows synergism is widely used for an effective control in corn.
iii. **Antagonistic effect:** The total effect of a combination is smaller than the effect of the most active component applied alone e.g. Combination of EPTC with 2,4-D, 2,4,5-T or dicamba have antagonistic responses in sorghum and giant foxtail. Similarly, chlorpropham and 2,4-D have antagonism. When simazine or atrazine is added to glyphosate solution and sprayed the glyphosate activity is reduced. This is due to the physical binding within the spray solution rather than from biological interactions within the plant.
iv. **Independent effect:** The total effect of a combination is equal to the effect of the most active component applied alone
v. **Enhancement effect:** The effect of a herbicide and non-toxic adjuvant applied in combination on a plant is said to have an enhancement effect if the response is greater than that obtained when the herbicide is used at the same rates without the adjuvant e.g. Mixing Ammonium sulphate with Glyphosate.

Herbicide-Moisture Interaction

Soil applied herbicides fail when there is a dry spell of 10-15 days after their application. Pre-emergence herbicides may be lost by photo-decomposition, volatilization and wind blowing while some amount of water is desirable to activate the soil applied herbicides, excess of it may leach the herbicide to the crop seed and root zone. This may injure the crops and on other side, results in poor weed control. Heavy showers may wash down herbicides from the foliage. Continuous wet weather may induce herbicide injury in certain crops by turning them highly succulent e.g. Maize plants are normally tolerant to Atrazine but they become susceptible in wet weather, particularly when air temperature is low. Extra succulence has been found to increase atrazine absorption and low temperature decreases its metabolism inside the plants. Quality of water used may also determine herbicide action. Dusty water reduces action of paraquat. Calcium chloride rich water reduces glyphosate phytotoxicity.

Herbicide-Insecticide Interaction

These chemicals are usually not harmful at recommended rates. The tolerance of plants to a herbicide may be altered in the presence of an insecticide and vice versa. The phyto-toxicity of monuron and diuron on cotton and oats is increased when applied with phorate. Phorate interacts antagonistically with trifluralin to increase cotton yield, by stimulating secondary roots in the zone of pesticide incorporation. Propanil interacts with certain carbamate and phosphate insecticides used as seed treatments on rice. But chlorinated hydrocarbon insecticides as seed treatment have not interacted with propanil. When propanil is applied at intervals between 7 and 56 days after carbofuron treatment, it results in greater injury to rice vegetatively.

Herbicide-Pathogens / Fungicides Interaction

Herbicides interact with fungicides also. In sterilized soil, chloroxuron is not causing any apparent injury to pea plants, while in the presence of *Rhizoctonia solani* in unsterilized soil it causes injury. Oxadiazon reduces the incidence of stem rot caused by the soil borne pathogen *Sclerotium rolfsii* L. in groundnut. Diuron and triazine which inhibit photosynthesis may make the plants more susceptible to tobacco mosaic virus. On the other hand, diuron may decrease the incidence of root rot in wheat.

Herbicide-Fertilizer Interaction

Herbicides have been found to interact with fertilizers in fields e.g., fast growing weeds that aregetting ample nitrogen show great susceptibility to 2,4-D, glyphosate than slow growing weeds onpoor fertility lands. The activity of glyphosate is increased when ammonium sulphate is tankmixed. Nitrogen invigorate (put life and energy in to) the meristematic activity in crops so much that they susceptible to herbicides. High rates of atrazine are more toxic to maize and sorghumwhen applied with high rates of phosphorus.

Herbicide-Microbes Interaction

Microorganisms play a major role in the persistence behaviour of herbicides in the soil. The soil Microorganisms have the capacity to detoxify and inactivate the herbicides present in the soil. Somegroups of herbicides more easily degrade through microbes than others. The difference lies in themolecular configuration of the herbicide. The microorganisms involved in herbicide degradationinclude bacteria, fungi, algae, moulds etc. Of these, bacteria predominates and include the membersof the genera Agrobacterium,Arthrobacter, Achromobacterium. Bacillus, Pseudomonas, treptomyces, Flavobacterium, Rhizobium etc. The fungi include those of thegenera Fusarium, Penicillium etc.

The students will be required to perform compatibility test for the common herbicides and fertilizers in use

Fertilizer	Herbicide						
	1	2	3	4	5	6	7

Interpretation

___

___

___

Exercise- 15

Study of Methods of Herbicide Application and Spraying Equipment's

A spray is defined as liquid discharged in particles and scattered as dispersed droplets. Sprayer is an appliance which atomizes the spray fluid which may be a suspension, an emulsion or a solution. Different sprayers for different purposes and conditions are available in the market. Knowledge of sprayers is essential for effective application of herbicide and their proper maintenance. There are various application methods viz. spraying or broadcasting for treating weeds with herbicides. It's important to choose the right method for your particular weed problem and the types of chemicals you are using. Different methods by which the herbicides are applied in the crop lands are

1. Soil Application of Herbicides:

i. Surface Application

Soil active herbicides are applied uniformly on the surface of the soil either by spraying or by broadcasting. The applied herbicides are either left undisturbed or incorporated in to the soil. Incorporation is done to prevent the volatilization and photo-decomposition of the herbicides. e.g. Fluchoralin – Left undisturbed under irrigated condition - Incorporated under rainfed condition.

ii. Subsurface Application

It is the application of herbicides in a concentrated band, about 7-10 cm below the soil surface for controlling perennial weeds. For this, special types of nozzles are introduced below the soil under the cover of a sweep hood. e.g.

carbamate herbicides to control *Cyperus rotundus*; Nitralin herbicide to control *Convolvulus arvensis.*

iii. Band Application

Application to a restricted band along the crop rows leaving an untreated band in the inter-rows. Later inter-rows are cultivated to remove the weeds. Saving in cost is possible here. For example when a 30 cm wide band of a herbicide applied over a crop row that were spaced 90 cm apart, then two-third cost is saved.

iv. Fumigation

Application of volatile chemicals in to confined spaces or in to the soil to produce gas that will destroy weed seeds is called fumigation. Herbicides used for fumigation are called as fumigants. These are good for killing perennial weeds and as well for eliminating weed seeds. e.g. Methyl bromide, Metham.

v. Herbigation

It is the application of herbicides with irrigation water both by surface and sprinkler systems. In India farmers apply fluchloralin for chillies and tomato, while in western countries application of EPTC with sprinkler irrigation water is very common in Lucerne.

2. Foliar application

i. Blanket Spray

It is the uniform application of herbicides to standing crops without considering the location of the crop. Only highly selective herbicides are used here e.g. Spraying 2,4-Ethyl Ester to rice three weeks after transplanting.

ii. Directed Spray

It is the application of herbicides on weeds in between rows of crops by directing the spray only on weeds avoiding the crop. This could be possible by use of protective shield or hood. For example, spraying of glyphosate in between rows of tapioca using hood to control *Cyperus rotundus.*

iii. Protected Spray

It is a method of applying non-selective herbicides on weeds by covering the crops which are wide spaced with polyethylene covers etc. This is expensive and laborious. However, farmers are using this technique for spraying glyphosate to control weeds in jasmine, cassava, banana.

iv. Spot treatment

It is usually done on small areas having serious weed infestation to kill it and to prevent its spread. Rope wick applicator and Herbicide glove are useful here.

Methods to treat shrubs and trees foliar spray

In foliar spraying, the herbicide is diluted with water or diesel at a specific rate, and sprayed over the foliage to point of run-off (until every leaf is wetted, but not dripping).

This method is most suited to shrubs, grasses and dense vines less than 6m tall. Advantages include quickness and economy. Disadvantages include the potential for spray drift and off-target damage.

Foliar spraying can be done a number of ways, including:

- Blanket spraying using a boom spray from a tractor
- A hose and handgun spraying solution from a herbicide tank
- A backpack spray unit
- With splatter guns (larger droplets at higher concentrations) for regrowth.

Basal Barking

This method involves mixing an oil-soluble herbicide in diesel and spraying the full circumference of the trunk or stem of the weed.

Basal bark spraying is suitable for:

- Thin-barked woody weeds
- Undesirable trees
- Saplings, regrowth, and multi-stemmed shrubs and trees.

Basal barking will usually destroy weeds as long as the bark is not wet or too thick for the diesel to penetrate.

Girdling

Girdling is often used to control trees or shrubs that have a single trunk. It involves cutting away a strip of bark several centimeters wide all the way around the trunk. The removed strip must be cut deep enough into the trunk to remove the vascular cambium, or inner bark, the thin layer of living tissue that moves sugars and other carbohydrates between areas of production (leaves), storage (roots), and growing points. This inner cambium layer also produces all new wood and bark.

Stem Injection

The stem injection method involves drilling or cutting through bark into the sapwood tissue of woody weeds and trees to transport the chemical throughout the weed. It is essential to apply the herbicide immediately (within 15 seconds of drilling or cutting), as stem injection relies on the active uptake and growth of the weed to move the herbicide through its tissue.

Drill and fill Method

The drill and fill method is used for trees and woody weeds with stems or trunks greater than 5 cm in circumference, and is also referred to as tree injection.

This method uses a battery-powered drill to drill downward-angled holes into the sapwood approximately 5cm apart. Using a backpack reservoir and syringe can deliver measured doses of herbicide solution.

Only trees and shrubs that can be safely left to die and rot, should be treated this way. If the tree or shrub is to be felled, allow it to die completely before felling.

Axe Cut Method

The axe cut method can be used for trees and woody weeds with stems or trunks greater than 5 cm in circumference. It involves cutting through the bark into the sapwood tissue in the trunk, and immediately placing herbicide into the cut. The aim is to reach the tissue layer just under the bark, which will transport the herbicide throughout the weed.

Using an axe or tomahawk, horizontal cuts are made into the sapwood around the circumference of the trunk at waist height. The axe is then leaned out to make a downward angled pocket to allow herbicide to pool. It's important not to entirely ringbark the trunk, as this will decrease the uptake of herbicide.

Cut Stump

The cut stump method involves cutting off the weed completely at its base (no higher than 15cm from the ground) using a chainsaw, axe, brushcutter or machete. A herbicide solution is then sprayed or painted onto the exposed surface of the cut stump, with the objective of destroying the stump and the root system.

It is essential that the herbicide solution is applied as soon as the trunk or stem is cut. A delay of more than 15 seconds for water-based herbicides and 1 minute for diesel-soluble herbicides will give poor results.

Cut and Swab

This method is similar to the cut stump method, but is suited to vines and multi-stemmed shrubs. Here, the weed stems are cut through completely, close to the ground. Herbicide is then applied immediately to the cut surface emerging from the ground, via spray or brush application.

Stem Scraper

Stem scraping is used for weeds and vines with aerial tubers. A sharp knife is used to scrape a very thin layer of bark from a 10cm section of stem. Herbicide is then immediately applied to the exposed, underlying green tissue.

Wick Applicators

This method consists of a wick or rope soaked in herbicide from a reservoir attached to a handle, or assisted with 12-volt pump equipment. The wetted wick is used to wipe or brush herbicide over the weed.

Equipments Used for Application of Herbicides Spray Equipments:

Herbicides are largely applied as spray. Several types of sprayers are available from small hand operated to large ground and aerial sprayers.

Components of Sprayers:

The important components are:

1. Pump
2. Power source
3. Tank
4. Agitator
5. Distribution system
6. Pressure gauge
7. Pressure regulator.

1. *Pump:*

Any spray liquid must be atomized before it leaves the spray nozzle. The pump provides the necessary pressure for this purpose.

Types of Pumps:

a. Air Compression or Pneumatic pumps:

These pumps force air into an air tight tank containing spray liquids thus moving the spray liquid under pressure through the nozzle for its atomization.

b. Hydraulic or Positive Displacement Pump:

These pumps take in a definite volume of spray liquid and force it through the delivery system under pressure. The pump differs in pressure they produce.

2. *Source of Power:*

It is needed to run the spray pumps. The source of power may be either

a) Manual
b) Traction
c) Motor
d) Tractor and air craft engines.

3. *Spray Tank:*

A sprayer may have either built in tank or a separate tank to carry spray liquid. The tank should be large enough to avoid frequent refilling but not unhandy to carry. The tank is provided with a large opening fitted with a strainer and cap to fill in the liquid. It is difficult to fill in liquid and clean the tank having small openings.

4. Agitator:

It may be either mechanical or hydraulic purpose, to keep liquid spray homogenous. Mechanical agitators may be of metal fan or rod etc. Hydraulic agitator consists of a pipe with several side holes and closed at its free end is placed in the tank and it is fed with spray liquid from the pump. From these holes the liquid emerses as jets to provide agitation to the whole body of the liquid. This is called as '**By pass system**'. Hydraulic agitation is not thorough but it is more convenient in power sprayers using on large tank size. Sprayer without agitator should not be used to apply pesticide emulsion and suspension.

5. Distribution System:

It includes

i. Nozzle
ii. Spray lance
iii. Spray boom
iv. Hose

i. Nozzle:

The function of spray nozzle is to break pressurized spray liquid into droplets for application to the target Nozzles are identified by

a) Droplet size
b) Delivery and
c) Spray pattern that they produce spray pattern is fixed for a herbicide work, eight kinds of spray nozzles are common e.g.

1. Flat fan
2. Solid cone
3. Flooding
4. Tripe action
5. Broadcast fan
6. Blast
7. Low volume
8. Centrifugal (Sprinkler rotary).

The "Flat fan" nozzles are available in two spray patterns viz. the tapped edge pattern and rectangular pattern. Tapped edge pattern to apply pre and post emergence herbicide broadcasting, while rectangular pattern for the pre emergence bank application of herbicides.

Solid cone nozzle produces medium size droplets. Good for pre and post emergence spray. Also used for surface application of herbicides which gives fan like spray.

Triple action nozzles-diameter of the sprays can be easily changed during operating to produce either coarse or fine spry.

Broad cast fan nozzles are used for spraying on unwanted vegetation, road side fence, rows etc. it gives wide coverage of 5 to 8 m with coarse droplets on emulsion to avoid drift.

Motorized sprayers blowers employ blast nozzles. These nozzles feed the spray liquid into the air steam to split it into droplets and carry the droplets by the velocity of the wind.

ii. Spray Lance:

It is brass rod or 90 cm length attached to a delivery hose pipe of sprayer and fitted to its free end with a replaceable nozzle. A herbicide spray lance is bent at its nozzle to form a goose neck. At the hose end it is provided with trigger mechanism to control flow liquid for specific purpose. The spray lance may be fitted with plastic shields to prevent chemical from drifting.

iii. Spray bar or Boom:

It consists of a horizontal pipe on which 2 or several nozzles are fitted and spaced at 50 cm apart. Boom length varies from 1 to 15 m. Short boom with 2-3 nozzles is used with manual sprayers, while longer ones with tractor sprayers. The main advantage of spray boom over spray lance is wide swath it covers in each trip of the sprayer over the field. Total width of land wetted by a boom can be adjusted to get either (i) Uniform spray (ii) Directed spray or (iii) Band spray

6. Pressure Regulator:

It is fitted to heavy duty sprayers and tractor driven sprayers so as to run the sprayers at constant pressure. Pressure gauge is provided to check pressure.

Types of Sprayers

A. Knapsack Sprayers:

Loaded on the back of worker during operations. Tanks may be plastic or metal. Common Knapsack sprayers are

i. Hydraulic
ii. Manual pneumatic and
iii. Motorized pneumatic.

i. Hydraulic Knapsac Sprayers:

Manually operated, tank capacity is 15 liters, mechanical or hydraulic agitation, worked with a hand lever to maintain constant pressure, particularly used for spot treatment small holding farmer and hand treatment. Equipped with a boom. It is good for blanket application.

Drawbacks:

- These sprayers are mounted on the back of man. One hand to the lever and the other to lance with a lance, one sprays 0.4 ha/day and with a boom 0.8 ha/day.
- It is high volume spray but low volume nozzles can be fitted.
- Spray potential is 12 kg/cm^2. It sprayed @ 3 to 4 kg cm^2 to prevent spray drift.

ii. Pneumatic or Compressed System Knapsac:

Do not require pumping during operation / spraying. The tank is pressurized after filling the liquid to 2/3rd capacity with a built in hand pump. Undesirable for weedicide spray pressure lower after some time spraying resulting into uneven spray. Tank cleaning is difficult. Used limited to spray on weeds in paddy and jute.

iii. Motorised Pneumatic Sprayers:

As a low volume sprayer suitable for spraying concentrated spray liquid. A blast of air flows through spraying jet of delivery hose and nozzle tube and ejects spray liquid in this blast. Air blast atomizes spray liquid in to fine droplets. Air acts as carrier. Faster the air is pressured, more the atomization. These sprayers are also used as blowers. The spray is inconsistent. Mist blower cause considerable loss of herbicide by winds. The main advantages of Knapsac blower are:

The main advantages of Knapsack Sprayer are:

1. Low volume spray. Loss of time in refilling tanks.
2. Portable working.
3. Fast spraying suited to post emergence translocated type herbicides as low volume Spraying is not so uniform with Knapsac blowers. Liquid – 60 liters/ha and swath 7 to 8 m.

B. Foot Sprayer Pedal Pump Sprayers:

Popularly applied for pesticide application operated with foot. It has provision of 1 – 2 long delivery hoses. Fitted with either lance or 2-6 nozzle booms. Its potential spray pressure is 17 to 21 kg/cm^2. Output with lance is 1 ha/day. It can spray high volume spray and covers more area.

C. Traction Pneumatic Sprayer:

Indian Institute of Sugarcane Research, Lucknow has developed bullock drawn sprayer with size nozzle boom that of powered from the wheels of the frame. It is efficient, easy to operate and simple in its construction. It uses two pneumatic pumps and develops maximum pressure of 2-8 cm^2 which his suited to minimize spray drift. Area covered 2-3 ha/day equipment.

C. *Tractor Mounted Sprayers:*

With spray pressure of 1.4 to 2.8 kg cm^2 and fitted with multi nozzle boom are very useful in herbicide application for large holding of farmers. Tractor mounted sprayer fitted with booms are used to spray road side vegetation. Tractor run sprayers have.

1. High uniformity of sprayers.
2. High working efficiency.
3. Full utilization of tractor during idle time.

A. *Aerial Sprayers:*

Herbicide application from air is limited to treat aquatic weeds like water hyacnth, paddy fields, large sugarcane plantation. Presence of obstacles like trees and diversified farming in India are bottle necks in its use.

As per Indian Standard Institute norms, on the basis of amount of solution required, the sprayers are classified as

S. No.	Class	Volume (l/ha)
i.	High volume (HV) or full coverage spray	>560
ii.	Medium Volume (MV) or semi low or semi- concentrated spray	56-560
iii.	Low volume (LV) or concentrated spray	5.6-56
iv.	Ultra Low Volume (ULV) spray	0.56-5.6
v.	Ultra-Ultra Low Volume (UULV) spray	<0.56

Sprays are also classified in the following ways

S.No.	Sprays	Symbols	Spray volume	
			Field crops	Trees and bushes
i.	High volume	HV	>600	>1000
ii.	Medium volume	MV	200-600	500-1000
iii.	Low volume	LV	50-200	200-500
iv.	Very low volume	VLV	5-50	50-200
v.	Ultra low volume	ULV	<5	<50

Droplet size plays a significant role in CPP application by minimizing environmental contamination. CPP sprays are generally classified according to droplet size. When drift is to be minimized, a medium or coarse spray is required irrespective of the volume applied.

Droplet size will influence coverage and drift. The nozzles typically used to apply CPPs produce droplets that vary in size to a great extent. Large droplets, which will facilitate in mitigating spray drift, may not provide good coverage. Very small droplets lack the momentum needed towards the target and are prone to drift under windy conditions. Flow rate of liquid (size of nozzle orifice), liquid pressure, physical changes to nozzle geometry and operation are the factors determining the range of droplets from a nozzle.

Classification of sprays according to droplet size

Volume medium diameter of droplet (μm)-1	Classification of droplet size
<50	Aerosol
51 - 100	Mist
101 - 200	Fine spray
201 - 400	Medium spray
>400	Coarse spray

The most widely used parameter of droplet size is volume medium diameter (Vmd) which is measured in micrometers (μm) . 1 mic = 1/1000 mm.

Maintenance and Cleaning of Sprayers Maintenance of Sprayers:

i. Use of clean water only.
ii. Use of the screen at the inlet spray.
iii. Use of metal object for cleaning the nozzles.
iv. Flush new sprayers before their use.
v. With Phenoxy herbicide use separate barrels of tanks if possible.
vi. Clean the sprayer thoroughly after each period of use.

Cleaning of Sprayers:

It is necessary to remove all residues of herbicides completely after spraying is completed. It is essential for prevention of following:

a) Damage to crop plants subsequently sprayed with different herbicides.
b) Undesirable action between herbicide residue and new herbicide used.
c) Corrosion of sprayer parts.

Procedure for Cleaning of Sprayers:

1. Remove and clean all screens and boom extensions with kerosene and a small brush.
2. Mix one box of detergent with 30 gallons of water in tank. Circulate through by pass system or 30 minutes and the drain out.
3. Replace the screens and the boom extensions.

4. Fill the tank 1/3 rd to ½ with one part of hose hold ammonia to 49 parts of water. Circulate this mixture through the pump and nozzles. Let the remaining solution stand over night and then run it over through the nozzle.
5. Flush with two tanks full of clean waters spraying through the boom with the nozzle removed

Exercise - 16

Calculation of Herbicide Doses

Requirement of a herbicide depends upon the type of herbicide, formulation, per cent of active ingredient, area to be sprayed and volume of water. Further, the method of calculation of herbicide dose will be different for aquatic situation (water body) in comparison to unit area of land.

Preparation of Herbicidal Sprays at Appropriate Concentrations

In order to achieve good results, it is not only necessary to use adequate quantities of recommended herbicides but the herbicidal sprays should be correctly prepared and properly applied. In this regard some important considerations are as follows:

1. Time of Application:

The time of application of different herbicides varies with the crop. The right time for spraying 2,4-D in wheat is at fully tillered stage. It is sprayed earlier, the ears will be abnormal and if sprayed later, weeds may not be killed. There are three times of application viz. pre-planning, pre-emergence and post-emergence application.

2. Concentration of herbicides:

It should be just as per the recommendations and if necessary repeated sprays may be given.

3. Formulation of Herbicides:

The herbicides are formulated in different ways to affect is solubility, volatility and toxicity to plats. The herbicides are formulated to be applied as solutions of water or oils, emulsions, wettable powders, granules and dusts.

4. Method of Application:

Different methods of application of herbicides are possible e.g. broad cast or blanket application, band application, placement, direct sprays by adjusting the height of the nozzle and spot treatment for small specified area.

5. Application Used their Care and Maintences:

Equipments used for application of herbicides may be sprayers power Knapsac of foot paddle for spraying liquids. Adusters for dispersing solids, fumigators and injectors for distributing material on the soil surface or in the soil. In order to use all these equipments efficiently proper lubrication, cleaning, checking of nuts, filters and hoses, calibration, using clean, nozzles proper cleaning and washing after use and proper storage in go down are necessary.

6. Calculating Proper Quantities of Herbicides:

The herbicides are available either in solid or liquid form. The label of the containers will read a.e = Acid equivalent or a.i. active ingredient for solids and g/lit a.e. or a.e. for liquids.

Acid Equivalent:

It refers to that part of a formulation that theoretically can be converted into acid. In this case the acid equivalent is given as the active ingredient or acid equivalent is given on label.

Active Ingradient:

It is that part of a chemical formulation which is directly responsible for herbicidal effect. Thus the commercial herbicide production is made up of two parts i.e. the effective part and the inert part. Since all the recommendations are made on the basis of a.e. or a.i. certain amount of calculations becomes necessary to find out the quantity of commercial product to be required for a given area. To calculate the weight of the commercial produce required, the formula used is:

$$\text{Weight of commercial material required} = \frac{\text{Weight of chemical to be applied a.i.}}{\text{\% of a.i. expressed as decimal}}$$

Example No. 1

If you buy a herbicide with 75% (0.75 a.i.) as diuron and want to apply two kg a.i. of Diuron per ha, then the quantity of commercial product required per ha will be

$$\frac{200}{0.75} = 2666 \text{ g of Diuron per ha.}$$

The 2666 g of the commercial product is required to add to the amount of water required to cover a hectare.

Example No. 2

Suppose Na salt of 2.4-D contains 80% a.i. and if 2 ½ kg of a.i. per ha is to be sprayed. The quantity of Na salt of 2,4-D will be required

$$\frac{2500}{0.80}=\frac{2500}{80}=\frac{100}{1}=3125\,g/ha=3.125kg/ha$$

Example No. 3

The Na salt of 2,4-D with 80% a.i. is to be sprayed on 3/10 of ha at 0.75 kg a.i. /ha. The quantity of Na salt required will be:

$$\frac{0.75}{0.80}\times\frac{30}{10}=0.2812kg=281.2g$$

Example No. 4

If in area of 1.5 ha is to be sprayed with simazine (50% WP) at 2 kg a.i./ha, the quantity of commercial herbicide required will be.

$$\frac{2}{0.50}\times\frac{3}{2}=6\,kg\,for\,1.5\,ha$$

Parts Per Million (PPM):

It refers to the number of parts by weight or volume of a constituent in 1,00,000 parts of final mixture by weight on volume. The stated conc. should tell whether PPM is measured by weight or volume. For herbicidal purposes, it usually refers to a given weight of a chemical in a given volume of spray.

Percent Concentration:

It is similar to ppm except that it is expressed as percentages, 1000 ppm is equivalent to 0.1% and 5000 pm to 0.5%. 1% solution (10000) ppm contains one g of 2,4-D (acid equivalent in 100 cm^3 of spray).

Exercise

a) Determine amount of isoproturon 75% WP for spraying 14.5 ha wheat crop (Given dose of isoproturon is 1.25/ha)

b) A farmer of Solan has a problem of isoproturon resistant *P. minor* in his field measuring 8000 m^2. Calculate the amount of Topik 10% for the said area (Recommended dose of clodinafop is 60 g/ha).

c) Calculate amount of 2,4-D amine salt required for spraying a pond 45 m in length, 25 m width and 5 m average depth infested with water hyacinth. Recommended dose of herbicide is 10 ppm.

Exercise - 17

Calibration of Spray Equipment's

Generally, herbicides are applied in the form of solution. Process of finding out the exact quantity of water required for spraying unit area is called calibration. Sprayer is the basic equipment used for application of herbicides. Proper application of herbicide depends upon the proper adjustment of all the basic components of a sprayer Therefore, for uniform spraying of herbicides with good efficiency, it is necessary to calibrate the amount of water to be applied, speed of walking, pressure to be maintained etc. The main aim of calibration is to adjust the application pressure and application speed (walking speed) to get the desired volume rate.

Accurate application of herbicides depends on the accurate calibration of spray pump. Calibration implies to adjust the sprayer to apply exact quantity of carrier (water) for spraying certain area under a set of conditions. Application rate is dependent on nozzle type, size of the nozzle orifice, spraying pressure, spraying speed etc. Care must be taken that speed and pressure should not vary from the calibration test to actual field spraying.

Procedure

1. Fill the spray tank to a desired level with clean water and note it.
2. Operate the spray pump and spray in a known size area.
3. Measure the amount of spray (water) applied.
4. Calculate the quantity of water required per hectare as below:

$$\text{Application rate}(\text{l}/\text{ha}) = \frac{\text{Amount of water applied}(\text{l}) \text{x} 10000}{\text{measured area sprayed}(\text{N}^2)}$$

5. Calculate the number of spray loads per hectare as follow:

$$\text{No. of spray loads / ha} = \frac{\text{Application rate} (l/ha)}{\text{Tank capacity of sprayer} (l)}$$

6. Calculation of commercial dose

Each herbicide carries a label. To calculate the weight of commercial product required, information on weight of chemical to be applied and its active ingredient content are required:

$$\text{Commercial dose} (kg/ha) = \frac{\text{Dose of herbicide ai to be applied} (kg/ha)}{\text{a. i. content in herbcide formulation (percentage expressed as decimal)}}$$

Example: A herbicide with 30% a.i. as pendimethalin. If 0.9 kg/ha of pendimethalin is to be applied then 0.9/0.3 = 3.0 kg/ha of commercial formulation of pendimethalin (Stomp 30 EC) is required.

7. Calculate the amount of herbicide to be mixed in each spray load as follow

$$\text{mount of herbicide per load} = \frac{\text{commercialdose} (kg/ha)}{\text{Number of loads per ha}}$$

Conclusions: What type of nozzle will you recommend to the farmer for uniform application of herbicides?

____

____

The area covered per hour can be calculated as shown under

$$\text{Area} (ha/hr) = \frac{\text{Walking speed} (km/hr) \text{ x m / km x spray width} (m)}{m^2 / ha}$$

For example if a person is walking at 1 km/hr covering a swath of 0.6 meters wide, the area covered/hr is

$$\frac{1 \times 100}{1000} \times 0.6 = 0.06 ha/hr$$

At this rate, it will take 16 hr 40 minutes to cover an area of one hectare. At a spray discharge rate of 30 l/hr, it would require a spray volume of 500 l/ha.

Exercise - 18

Types of Sprayers and Parts

Spray Equipment is Chosen According to

- Discharge requirement
- Droplet size

Select the spray equipment according to the required discharge as well as droplet size as the efficacy of applied herbicide depends upon these parameters.

Types of Sprayers

Depending upon the type of force required, sprayers are divided into two groups:

1. Manually operated

i. Compressed Air Sprayer

ii. Hydraulic Sprayer

2. Power Operated

i. Compressed Air or Pneumatic Sprayer

A. Domestic Sprayer

These are only used to spray against flies, mosquitoes and in kitchen gardening, hospitals, poultry farms etc.

B. Hand compressing sprayers

Most common is Knap-sack sprayer. This pump is very ideal and economical for spraying pesticides including herbicides, insecticides, fungicides, etc.

C. *Pressure Retaining or Battery Sprayer*

Pressure is maintained throughout the spraying time.

ii. Hydraulic Sprayers

A. *Pedal Pump (Foot Sprayer)*

This pump is not suitable for spraying herbicides. It can be used for spraying insecticides and fungicides especially in orchards and plantations.

B. *Power Operated Sprayers*

In these sprayers, pressure is generated mechanically with diesel or petrol engines with varying HP. These consist of a power generating pump, suction chamber, delivery as well as discharge lines etc. The spray fluid is blown out by means of an intensive air current generated by the blower. The spray liquid blows through PVC tube and reach the nozzle where the speedy air from blower, blasts the liquid into small droplets, which can be measured in microns. The capacity of these power sprayers may vary from 15 litre to more than 500 litre. These are very ideal equipments for spraying foliar uptake herbicides.

Parts of Knapsack Sprayer

The knowledge of various parts of a sprayer and their functions is necessary for efficient operation, maintenance and repair. Major parts of knap sack sprayer are given below:-

a) Tank/Container

It holds the spray fluid. The total capacity of tank of knap-sack sprayer is 15 litres.

b) Pressure Pump

It is required to build up pressure in the system and is necessary for atomization of spray fluid into smaller droplets.

c) Discharge Line/Delivery Line

Through this liquid comes out of the sprayer with force after converting to mist form. It consists of the following parts:-

I. Delivery hose: A flexible tubing of rubber or plastic whose one end is attached to the tank / container.
II. Lance: A metal tube attached to the other end of hose and up to nozzle.
III. Cut off/Shut off: Generally fixed at the junction of the delivery hose and lance to cut off the flow of liquid when spraying is suspended.
IV. Nozzles: These are of three types

i. Flat fan
ii. Flood jet
iii. Cone

d) Handle

It is moved up and down to generate pressure.

e) Agitators

The active materials in any spray fluid are not always in solution, but are held in suspension form and consequently the material should be kept well mixed in order to secure an even distribution of herbicides.

f) Filters/Strainers

The function of filters is to prevent solid particles for clogging the system.

g) Washers and Gaskets

These are present at all joints to prevent leakage.

h) Shoulder straps and hooks

Their function is to hold the sprayer on shoulder of the man engaged for spraying.

Manual operated Knap sack sprayer

Power operated Knap sack sprayer

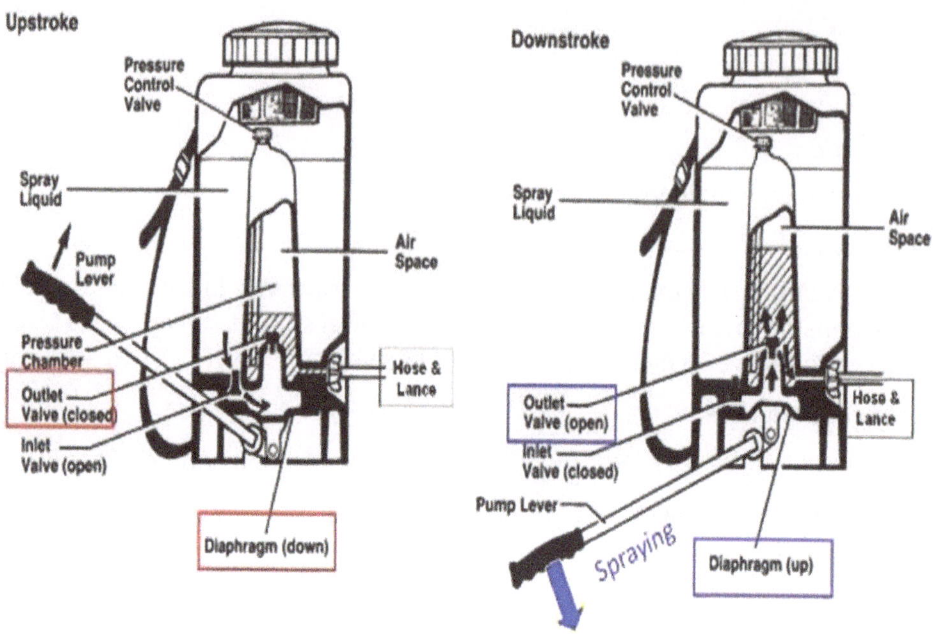
Upstroke
Pressure Control Valve
Spray Liquid
Pump Lever
Air Space
Pressure Chamber
Outlet Valve (closed)
Inlet Valve (open)
Hose & Lance
Diaphragm (down)
Downstroke
Pressure Control Valve
Spray Liquid
Air Space
Outlet Valve (open)
Inlet Valve (closed)
Hose & Lance
Pump Lever
Spraying
Diaphragm (up)

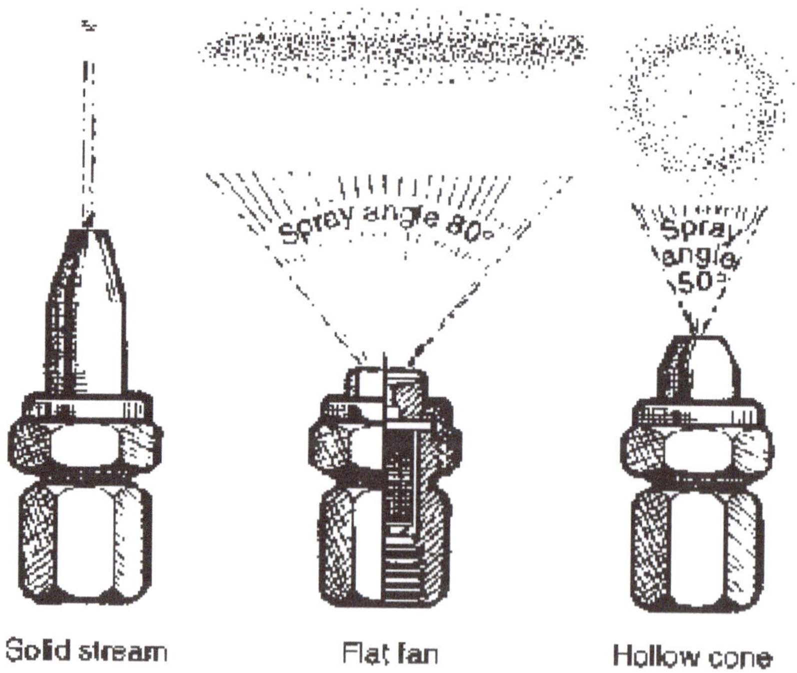
Spray angle 80°
Spray angle 50°
Solid stream
Flat fan
Hollow cone

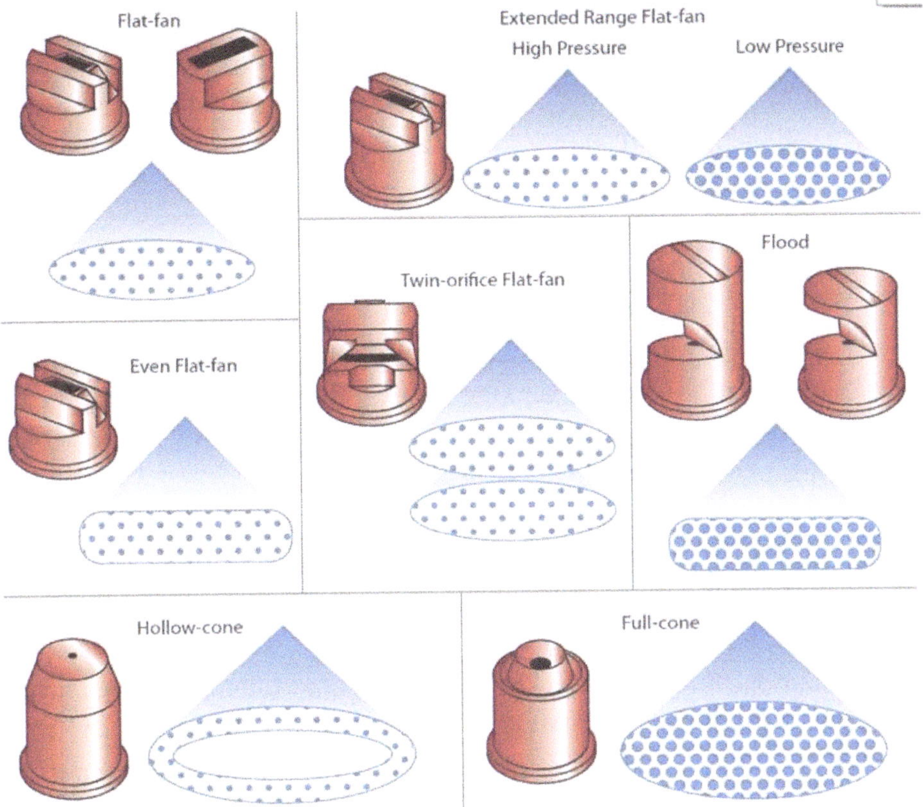

Types of Sprayer Nozzle and Spray Pattern

Spray Technology

- Before starting the spraying operation check the nozzle by spraying plain water on a pucca floor.
- If wetting of the floor is not uniform, change the nozzle or nozzle tip.
- The person who has done calibration, should be deputed for spray in the field.
- Start spraying from one corner of the field in a band and second band should be parallel to the first with slight overlapping.
- Keep the nozzle at the height of 50 cm from the ground or canopy level.
- Spray should be as uniform as possible. The left over solution should not be re sprayed but must be discarded on a barren land.
- Do not move the nozzle to and fro.
- Always spray at right angle to the direction of wind and do not spray along or across the direction of wind.

- Postpone spray, if wind velocity is high (>10 km per hour).
- Wear full sleeve shirt, trouser, shoes, hand gloves and gas mask, while spraying.
- Do not spray with empty stomach and also do not eat while spraying.
- After spray operation is complete, take bath with soap and change your clothes.

How to Prepare Spray Fluid?

- Thoroughly mix the calculated dose of herbicide in small quantity of water and then make the volume of solution (spray fluid) upto calibrated amount in a big container.
- Use hand gloves and gas masks while preparing the solution and at the time of its application in the field.
- Stir the spray fluid thoroughly before each filling.
- Alternatively, stock solutions (number of litres equivalent to number of spray pumps required for spraying one hectare) can be prepared in a small container.
- Pour one litre of stock solution into the pump and make the volume up to 15 litre by adding water.
- Put one or two litres of water before pouring stock solution in the container.

Exercise - 19

Methods of Herbicidal Application and Necessary Precautions

Effective and efficient weed control depend on selection of proper method of application. Herbicides are applied to the soil (Soil application) or to the foliage (foliar application) depending on the properties of herbicide, mode of action and selectivity. Environmental factors, convenience and cost are the other factors that influence the choice of a correct method of application. An improper method of application can result in poor weed control and / or severe crop injury.

Methods of Herbicidal Application:

The herbicidal application can be grouped in the following ways-

1. On the Basis of time of Application

i. Pre-planting incorporation,
ii. Pre-emergence application,
iii. Post- emergence application, and
iv. Lay-by application.

2. On the basis of Method of Application-

A) Foliage application

i. Blanket application
ii. Directed application
iii. Protected application
iv. Spot treatment

B) Soil Application

i. Surface application
ii. Sub surface application
iii. Band application
iv. Soil fumigation

The choice of a method to be effective for weed control depends upon the nature and properties of herbicide and the characteristics of target plants. Herbicides, in natural state, may be solid or liquid, volatile or non-volatile, soluble or no soluble. Herbicides are formulated in soluble powders, wettable powders or in emulsifiable forms. To increase the biological effectiveness of herbicide adjuvants are added. These adjuvant modify the physical properties of herbicide and make it fit for easy application, long storage and biologically effective.

Precautions to be taken while handling herbicides:

i. Read the label on each container before using the contents and strictly follow the instructions there in.
ii. Destroy the empty containers by burying them at least 18 inches deep in an isolated area away from water resources.
iii. Apply the herbicides only at the time specified on the label. It is very necessary to observe the recommended intervals between treatment and pasturing/ harvesting of the crop.
iv. Always consult the product label or technical bulletin before applying the chemical.
v. Wear goggles, rubber gloves and other clothing as recommended on the label.
vi. Guard against possible injury to nearby susceptible plants.
vii. The herbicide should be kept in a safe place (away from seeds, fertilizers, edible grains, cattle feed etc.) where children and other unauthorized person do not have access.
viii. The herbicides should never be stored on the farm and should be used just after purchase.
ix. Never use the herbicide from unlabeled containers or kept in such a container whose instructions have been mutilated.

Precautions to be taken during preparing solution:

i. Read the available literature on or with the container and act accordingly
ii. Always make the solutions in glass or plastics containers.
iii. The quantity of solution should be fixed according to the efficiency of sprayer used for the purpose.
iv. Add soap emulsion, or other adhesive substances to increase the viscosity of the spray material.

v. Care should be taken that during preparing the solution no part of the body may have direct contact with the herbicide. During this period the person using herbicide should wear goggles, rubber gloves and mask and adopt other protective measures.

vi. Always have soap and fresh water in the place of preparing solution.

Necessary recautions during use:

i. Act as per instructions on the label

ii. Never try to identify the herbicides by feel or smell.

iii. Stop spraying operations at high wind speed, intense heat or when rain is just expected.

iv. Never try to blow clogged nozzle by mouth pressure.

v. Clean and wash sprayers carefully before and after use.

vi. Never work alone while handling the herbicides as well as their application.

vii. Do not eat, drink, smoke or rub your eyes with hands during spray.

viii.Avoid sprays reaching the nearby water bodies.

Reasons for phytotoxicity symptoms of herbicides on crops.

All the recommended herbicides are very safe for respective crops in which they are recommended provided they are used according to the recommended technology. Even recommended herbicides may show phytotoxicity symptoms under certain situation such as

1. Wrong time of application: Application of herbicides too early than its recommended time may cause yellowing and necrosis of crop. Application of 2,4-D on very young crop may cause tubular leaves and deformed wheat ears.

2. Use of over dose: Application of higher than recommended dose is also responsible for stunting, yellowing and necrotic growth of crop plants.

3. Adoption of wrong application technology: Improper application of herbicides not only harm the present crop but may also prove harmful for the succeeding crops due to residual toxicity.

4. Spray drift: Due to speedy winds on the day of spray, adjoining crops may be damaged.

5. Adverse climatic conditions: Heavy rains immediately after the application of herbicides, little root uptake of herbicides may even damage the tolerant crops.

6. Due to non-adoption of recommended technology: With the use of unapproved herbicides, crop is damaged. For instance, wheat crop may be damaged with the use of metribuzin as pre emergence but crop remain unaffected when it is applied as post emergence to wheat. Germination of wheat crop is adversely affected when seeds come in contact with triallate treated soil zone.

Exercise - 20

Classification of Herbicides

Introduction

Any chemical used for weed control is called as weedicide or herbicide. Herbicides belonging to one chemical family tend to have similar modes of action on plants and behaviour in soils. Therefore, to have systematic understanding of possible behaviour of new herbicides belonging to a particular group or family, the classification of herbicides is essential.

Objectives

To become familiar with different groups or families of herbicides.

Classification of Herbicides

1. Based on Chemical Structure

Inorganic herbicides - Inorganic herbicides do not contain carbon atoms in their molecules, For example, Arsenic acid, sulphuric acid, sodium arsenate, sodium chlorate, borax, copper sulphate.

Organic herbicides - Organic herbicides contain carbon atoms in their molecules. They may be oils or non oils. Majority of the present day herbicides are organic compounds which are non oils.

2. Based on Selectivity

Selective : Selective herbicides kill only targete plants or weeds while crops are not affected eg. Simazine, altrazine, 2,4-D, MCPA, butachlor, alachlor, fluchloralin etc.

Non-Selective : Non-selective herbicides kill all vegetations that they come in contact with irrespective of whether it is a crop or weed, e.g. Paraquat and diquat.

2. Based on Translocation

Systemic – Systemic herbicides move within the plant either through xylem or phloem. Most of the herbicides are selective at recommended rates, Example, Altrazine, simazine, 2,4-D etc. **Contact** herbicides kill plants when they come in contact with plants. They kill the part of the plant that is in contact with herbicide s eg. Paraquat, diaquat.

Classification of Herbicides

Common name	Trade name	Structural formula
1. Aliphatic acids		
Dalapon	Dowpon, Tofapon	2,2-dichloro propionic and
TCA	Varitox	Trichloro acetic acid
2. Anilides and amides		
Alachlor	Lasso	2-chloro 2,6-diethyl N (methoxy methyl acetanilide)
Butachlor	Machete	N- (Butoxymethyl)-2 chloro-21,61 ethyl acetanilide
Propanil	Stam F-34, Rogue	3-4-dichloro propionanilide
3. Anilines and nitro-phenols		
Trifluralin	Treflan	2,6-dinitro-NN-dipropyl-4-trifluralin methylaniline
Fluchloralin	Basalin	N-propyl-N-(2-chloroethyl)-2,6-dinitro-trifluroethyl)-2-6 dinitro-triflurolin ethyl-aniline
Nitrofen	TOK E-25	2,4-dichlorophenyl nitrophenyl ether
4. Arsenicals		
DSMA	Ansar 184	Disodium methyl arsenate
MSMA	Ansar 529	Monosodium methyl arsenate

5. Benzoics and Phenyl acetic Acid		
Dicamba	Banvel-D	2,3,6-trichlorophenyl acetic acid
Oxyfluorfen	Goal	2-chloro-1 (3 ethoxy-4-nitrophenoxy)-4 (trifluro methyl) benzene
6. Carbamates and thio carbamates		
Barban	Carbyne	4chloro-2-butynyl m-chloro-carbanilate
EPTC	EPTAM	s-ethyl diprophyl thiocarnbamate
Glyphosate	Round up	N (Phosphono methyl) glycine
7. Heterocyclic compound		
Diquat	Reglone	1,1-ethylene-2,2 bi pyridylium dibromide
Paraquat	Gramoxone	1,1, dimethyl-4) 4-bipyridy liumdibromide
Altrazine	Atrataf	2-chloro-4 (ethyl amino)-6 (isopropyl amino)-S-triazine
Simazine	Gesatop	2,chloro-4,6 bis (ethyl amino)-Striazine
Metribuzin	Sencor	4-amino-6-tert-butyl 3 (methyl-thio) s-triazine-5 (4+1) one

8. Hormone type herbicides		
A. Phenoxy acetic acids		
2,4-D	Weedox, weedone	2,4-dichloro phenoxy acetic acid
2,4-5-T	Brush killer	2,4,5-trichloro phenoxy acetic acid
B. α phenoxy propionic acids		
Dichlorprop	2,4-DP	α-(2,4-dichlorophenoxy) propionic acid
Silvex	Fenoprop	2-(2,4,5-trichloro phenoxy) propionic acid

Mecoprop C. r – phenoxy butyric acids	MCPP	(4-chloro-2-methyl phenoxy) propionic acid
2,4-DB	Butaxone	(2,4-dichloro phenoxy) butyric acid
MCPB	Tropotox	(4-dichloro-2-methyl phenoxy) butyric acid
9. Substituted ureas		
Diuron	Karmex	3-(3,4-dichloro phenyl) dimethyl urea
Fenuron	Urab	1,1-dimethyl-3-phenyl urea mon
Monuron	Telvar	(trichloro acetate) 3(P-chlorophenyl)-1,1-dimethyl
Linuron	Loxox	Urea 3 (3,4-dichloro phenyl)-methoxy

Exercise - 21

Application of Pre Plant Herbicides in the Field

Introduction

Herbicides are the chemicals used to kill weeds. They are selective to crops when they are applied at recommended dose by proper method. For the effective and safe use of herbicides, time of application, selection of proper equipment, technique and method, knowledge of nature of chemical and environmental conditions are very important. A herbicide treatment made at any time before the crop is planted is called a pre plant treatment. Pre plant treatments are of two types (a) Pre plant desiccation and (a) Pre plant incorporation. In desiccation, herbicides are applied to destroy the standing vegetation as an aid to seed bed preparation. Whilc in prc plant incorporation. method, herbicides are mixed in weed free seed bed to obtain residual control of weeds during crop season.

Objective

To ensure proper application of pre plant herbicides in the field.

Materials

Knapsack sprayer, crop field, herbicide, water measuring scale etc.

Procedure

1. Selection of proper herbicide.
2. *Equipment :* according to area to be sprayed and method of application, choose an appropriate equipment. For spray, knapsack sprayers are suitable.

3. *Calibration* : Calculate the spray volume, walking speed, discharge rate and time of spray through calibration.
2. Compute the amount of herbicide needed.
3. Put on hand gloves, eye glass and tie cloth around the mouth.
4. Measure the amount of herbicide and put it in a bucket. Slowly pour some water in the bucket and stirr with a stick to obtaine homogeneous solution.
5. Add this solution to the water container and make up the desired volume.
6. Undertake the spraying operation.

Results

Collect and systematically note down the information.

Table 21.1 : Information required for herbicide application

Particulars	Description
A. About Herbicide	
1. Trade tame	
2. Quantity of formulated product for given area	
3. Stage of application	
B. About Requirement	
1. Name of equipment	
2. Working condition	
3. Type of nozzle	
4. Walking speed	
5. Spray volume for given area	

Exercise - 22

Application of Pre Emergence and Post Emergence Herbicides in Crop Field Introduction

Pre emergence treatment : Application of herbicides soon after planting of a crop is called pre emergence treatment. The soil should be adequately moist to induce weed seed germination and move the herbicide 1-1.5 cm deep into the soil in a uniform film.

Post emergence treatment : Application of herbicides after the emergence of both the crops and the weeds. When the weeds grow before the crop plants have emerged through the soil, the herbicidal application is called as **early post emergence treatment**. Herbicides used for these purpose are usually non-residual types.

Objective

To ensure proper application of pre and post emergence application of herbicide in the field.

Materials

Knapsack sprayer, crop field, herbicides, water, measuring scale etc.

Procedure

1. Selection of proper herbicide.
2. Measure the cropped area.
3. Calibrate the knapsack sprayer as explained in exercise 13.
4. Compute the herbicide needed.

5. Put on hand gloves, eye glass and a cloth around the mouth.
6. Measure the amount of herbicide and put it in a bucket and stir well.
7. Add this solution to the water container of sprayer and make the desired volume.
8. Undertake spraying operation.

Results

Collect and systematically note down the information.

Table 22.1: Information required for herbicide application

Particulars	Description
About Herbicide	
1. Trade tame	
2. Quantity of formulated product for a given area	
3. Stage of application	
A. About Requirement	
1. Name of equipment	
2. Working condition	
3. Type of nozzle	
4. Walking speed	
5. Spray volume for given area	

Exercise - 23

Application of Herbicides Under Aquatic Conditions

Introduction

Aquatic weeds are those unwanted plants which grow in water and complete at least a part of their life cycle in water. Aquatic weeds are of two types i.e. algae and hydrophytes. They cause harmful effects in several ways. Therefore, it is necessary to control aquatic weeds. Herbicides can be effectively used to control those aquatic weeds which are hard to mechanical or other control methods.

Objective

To control aquatic weeds.

Materials

Boat mounted power sprayers, herbicide, water.

Procedure

For proper application of herbicides in aquatic situation, following steps should be followed.

1. Survey the nature of weeds in pond or lake.
2. Selection of proper herbicide and time of application.
3. Choose an appropriate spray equipment.
4. Prepare solution of herbicide and spray it uniformly.
5. Use granular formulation of herbicides for control of submerged aquatic weeds.

Problem 1

1. Write down names of predominant aquatic weeds
2. Which herbicides are used to control aquatic weeds.

Exercise-24

Study of Phytotoxicity Symptoms of Herbicides in Different Crops

Introduction

Application of herbicides on field crops results some phytotoxicity symptoms. Higher doses of herbicides are also responsible for phytotoxicity.While conducting herbicidal spray utmost care should be taken regarding spray concentration and drift hazard.

Objective

To study the phytotoxicity symptoms of herbicides on crops.

Materials Required

Sprayer, water, 2,4-D, paraquat, bucket, measuring cylinder etc.

Procedure

1. Carry out spray operation in standing wheat crop with 2,4-D recommended dose.
2. Spray paraquat on Lucerne in *Cuscutta* infested field or blanket spray.

Observations

Phytotoxicity symptoms in wheat after a week

Phytotoxicity symptoms in lucerne after a week

Exercise - 25

Study of English Name, Scientific Name and Family of Important Weeds

S. No.	English Name	Scientific Name	Family
Kharif Weeds			
1.	Goose grass	*Acrachne racemosa*	Poaceae
2.	Running grass	*Brachiaria reptans*	Poaceae
3.	Wrinkle grass	*Ischaemum rugosum*	Poaceae
4.	Sandbur	*Cenchrus biflorus*	Poaceae
5.	Crow foot grass	*Dactyloctenium aegyptium*	Poaceae
6.	Crab grass	*Digitaria sanguinalis*	Poaceae
7.	Indian goose grass	*Eleusine indica*	Poaceae
8.	Red sprangletop	*Leptochloa chinensis*	Poaceae
9.	Water grass	*Echinochloa colona*	Poaceae
10.	Barnyard grass	*Echinochloa crusgalli*	Poaceae
11.	Love grass	*Eragrostis pilosa*	Poaceae
12.	Feather love grass	*Eragrostis tenella*	Poaceae
13.	Knot grass	*Paspalum distichum*	Poaceae
14.	Seashore paspalum	*Paspalum vaginatum*	Poaceae
15.	Rough bristle foxtail	*Setaria verticillata*	Poaceae
16.	Green foxtail	*Setaria viridis*	Poaceae

S. No.	English Name	Scientific Name	Family
17.	Hedgehog sedge	*Cyperus compressus*	Cyperaceae
18.	Rice flat sedge	*Cyperus iria*	Cyperaceae
19.	Erect horse weed	*Conyza stricta*	Astereceae
20.	Jimson weed	*Datura stramonium*	Solanaceae
21.	Smooth pigweed	*Amaranthus viridis*	Amaranthaceae
22.	Spiny pigweed	*Amaranthus spinosus*	Amaranthaceae
23.	False amaranth	*Digera arvensis*	Amaranthaceae
24.	False amaranth	*Digera muricata*	Amaranthaceae
25	Creeping chaff weed	*Alternanthera sessilis*	Amaranthaceae
26.	Alligator weed	*Alternanthera philoxeroides*	Amaranthaceae
27.	Blistering ammannia	*Ammannia baccifera*	Lythraceae
28.	Pink node flower	*Caesulia axillaris*	Astereceae
29.	Day flower	*Commelina benghalensis*	Commelinaceae
30.	Spreading day flower	*Commelina diffusa*	Commelinaceae
31.	Slender day flower	*Commelina erecta*	Commelinaceae
32.	Wild jute	*Corchorus tridens*	Tiliaceae
33.	Pill pod spurge	*Euphorbia hirta*	Euphorbiaceae
34.	Painted spurge	*Euphorbia heterophylla*	Euphorbiaceae
35.	Petty spurge	*Euphorbia prostrate*	Euphorbiaceae
36.	Shrubby spurge	*Euphorbia microphylla*	Euphorbiaceae
37.	Ground cherry	*Physalis minima*	Solanaceae
38.	Cockle bur	*Xanthium strumarium*	Astereceae
39.	Indian turnsole/Devil weed	*Heliotropium indicum*	Boraginaceae
40.	Black nightshade	*Solanum nigrum*	Solanaceae
41.	Green carpet weed	*Mollugo verticillata*	Molluginaceae
42.	Carpet weed	*Mollugo pentaphylla*	Molluginaceae
43.	False Daisy	*Eclipta alba*	Astereceae
44.	Yellow spider flower	*Cleome viscose*	Brassicaceae
45.	Purselane	*Portulaca oleracea*	Portulaceae
46.	Desert horse purslane	*Trianthema portulacastrum*	Aizoaceae
47.	Puncture vine	*Tribulus terrestris*	Zygophylaceae
48.	Gulf leaf flower	*Phyllanthus niruri*	Euphorbiaceae
49.	Indian sorrel	*Oxalis corniculata*	Oxalidaceae

S. No.	English Name	Scientific Name	Family
50.	Rattle pod	*Crotalaria medicaginea*	Fabaceae
51.	Coffee weed/ Sickle pod	*Cassia tora*	Fabaceae
52.	Creeping indigo	*Indigofera linnaei*	Fabaceae
53.	Tiger foot morning glory	*Ipomoea pestigridis*	Convolvulaceae
54.	Blue morning glory	*Ipomoea nil*	Convolvulaceae
55.	Wild carrot weed	*Parthenium hysterophorus*	Astereceae
56.	Tridex daisy	*Tridax procumbens*	Astereceae
57.	Spiny burr grass	*Cenchrus longispinus*	Poaceae
58.	Hair sedge	*Bulbostylis barbata*	Cyperaceae
59.	Grasslike fimbry	*Fimbristylis tenera*	Cyperaceae
60.	Common sedge	*Cyperus difformis*	Cyperaceae
61.	Yellow berried night shade	*Solanum xanthocarpum*	Solanaceae
62.	Golden daisy	*Vicoa indica*	Astereceae
63.	Worm weed	*Artemisia scoparia*	Astereceae
64.	Burn mouth vine	*Rhyncosia minima*	Fabaceae
65.	Wild melon	*Cucumis callosus*	Cucurbitaceae
66.	Globe amaranth	*Gomphrena celosioides*	Amaranthaceae
67.	Sun spurge	*Euphorbia helioscopia*	Euphorbiaceae
68.	Chicken weed	*Euphorbia thymifolia*	Euphorbiaceae
69.	Common spurge	*Croton sparsiflorus*	Euphorbiaceae
70.	Purslane	*Portuluca grundiflora*	Portulaceae
71.	Purslane	*Porlulaca quadrifida*	Portulaceae
72.	Pink wood sorrel	*Oxalis maritiana*	Oxalidaceae
73.	Little ironweed	*Vernonia cinerea*	Asteraceae
74.	Sand herbage	*Gisekia pharnaceoides*	Molluginaceae
75.	Wild buckweed	*Polygonum glabrum*	Polygonaceae
76.	Small knotweed	*Polygonum plebeium*	Polygonaceae
77.	Coffee senna	*Cassia occidentalis*	Fabaceae
78.	Arrowhead	*Sagittaria sagittifolia*	Alismaracese
79.	Frogfruit	*Lipia nodiflora*	Verbenaceae
80.	Hairy slitwort	*Lindernia ciliate*	Serofulariaceae

S. No.	English Name	Scientific Name	Family
81.	Purple spikerush	*Eleocharis atropurpurea*	Cyperaceae
82.	Green foxtail	*Setaria glauca*	Poaceae
		Rabi Weeds	
1.	Winter wild oat	*Avena ludoviciana*	Poaceae
2.	Wild oat	*Avena fatua*	Poaceae
3.	Sweet grass	*Poa annua*	Poaceae
4.	Beard grass	*Polypogon monspeliensis*	Poaceae
5.	Poison rye grass	*Lolium temulentum*	Poaceae
6.	Canary grass	*Phalaris minor*	Poaceae
7.	Wild onion	*Asphodelus tenuifolius*	Liliaceae
8.	Wild mustard	*Sisymbrium irio*	Brassicaceae
9.	Barrel clover	*Medicago truncatula*	Fabaceae
10.	California bur clover	*Medicago polymorpha*	Fabaceae
11.	Toothed bur clover	*Medicago denticulata*	Fabaceae
12.	Wild fenugreek	*Trigonella polycerata*	Fabaceae
13.	Common lamb squarter	*Chenopodium album*	Chenopodiaceae
14.	Nettle leaf	*Chenopodium murale*	Chenopodiaceae
15.	Green field-Speedwell	*Veronica agrestis*	Scropulariaceae
16.	Stone seed	*Lithospermum arvense*	Boraginaceae
17.	Canaigre dock	*Rumex hymenosepalus*	Polygonaceae
18.	Sour dock	*Rumex dentatus*	Polygonaceae
19.	Dock/Sorrel	*Rumex spinosus*	Polygonaceae
20.	Blue daisy	*Cichorium intybus*	Asteraceae
21.	Wild safflower	*Carthamus oxyacantha*	Asteraceae
22.	Maxican poppy	*Argemone mexicana*	Papaveraceae
23.	Perennial saw thistle	*Sonchus arvensis*	Asteraceae
24.	Little mellow	*Malva parviflora*	Malvaceae
25.	Meadow pea	*Lathyrus aphacaora*	Fabaceae

S. No.	English Name	Scientific Name	Family
26.	Grass pea	*Lathyru sativus*	Fabaceae
27.	Blue pimpernel	*Anagallis arvensis*	Primulaceae
28.	Purple cud weed	*Gnaphalium purpureum*	Asteraceae
29.	Hairy vetch	*vicia hirsute*	Fabaceae
30.	Vetch	*Vicia Sativa*	Fabaceae
31.	Forked catchfly	*Silene conoidea*	Caryophyllaceae
32.	Cow cockle	*Saponaria vaccaria*	Caryophyllaceae
33.	Yellow sweet clover	*Melilotus indicus*	Fabaceae
34.	White sweet clover	*Melilotus alba*	Fabaceae
35.	Fumatory	*Fumaria parviflora*	Fumariaceae
36.	Garden cress	*Coronopus didymus*	Brassicaceae
37.	Corn spurry	*Spergula arvensis*	Caryophyllaceae
38.	Cutleaf evening primrose	*Oenothera laciniata*	Onagraceae
39.	Cursed buttercup	*Ranunculus sceleratus*	Ranunculaceae
40.	Rough seed buttercup	*Ranunculus muricatus*	Ranunculaceae
41.	Corn Spurge	*Euphorbia segetalis*	Euphorbiaceae
42.	Chickweed	*Stellaria media*	Caryophyllaceae
43.	Wild dog flower	*Antirrhinum orontium*	Scrophulariaceae
44.	Catch weed	*Galium aparine*	Rubiaceae
		Perennial Weeds	
1.	Jhonson grass	*Sorghum halepense*	Poaceae
2.	Tiger grass	*Saccharum spontaneum*	Poaceae
3.	Bermuda grass	*Cynodon dactylon*	Poaceae
4.	Nut grass	*Cyperus rotundus*	Cyperaceae
5.	Field bind weed	*Convolvulus arvensis*	Convolvulaceae
6.	Blush morning glory	*Ipomoea carnea*	Convolvulaceae
7.	Prickly pear	*Opuntia dillenii*	Cactaceae
8.	Canada thistle	*Cirsium arvense*	Astereceae
9.	Kidney weeds	*Dichondra repens*	Convolvulaceae

S. No.	English Name	Scientific Name	Family
10.	Maddar root	*Calotropis procera*	Asolepiadaceae
11.	Lantana	*Lantana camara*	Verbinaceae
12.	Goat weed	*Ageratum conyzoides*	Astereceae
13.	Halfa grass	*Desmostachya bipinnata*	Poaceae
14.	Congo grass	*Imperata cylindrical*	Poaceae
15.	Marvel grass	*Dichanthium annulatum*	Poaceae
16.	Spiderling	*Boerhavia diffusa*	Nyctaginaceae
17.	Bold leaf launeae	*Launaea nudicaulis*	Verbinaceae
18.	Alyce clover	*Alysicarpus vaginalis*	Fabaceae
19.	Spanish needle	*Bidens pilosa*	Asteraceae
20.	Dallis grass	*Paspalum dilatatum*	Poaceae
21.	Caesar's weed	*Urena lobata*	Malvaceae
Parasitic Weeds			
1.	Loranthus	*Loranthus micranthus*	Loranthaceae
2.	Dodder	*Cuscuta reflexa*	Convolvulaceae
3.	Dodder	*Cuscuta chinensis*	Convolvulaceae
4.	Dodder	*Cuscuta europaea*	Convolvulaceae
5.	Witch weed	*Striga lutea*	Scropulariaceae
6.	Egyptian broomrape	*Orobanche aegyptiaca*	Orabanchaceae
7.	Broomrape	*Orobanche cernua*	Orabanchaceae
8.	Hemp broomrape	*Orobanche ramose*	Orabanchaceae
Aquatic Weeds			
1.	Water hyacinth	*Eichhornia crassipes*	Pontederiaceae
2.	White water lily	*Nymphaea alba*	Nymphaeaceae
3.	Water lettuce	*Pistia stratiotes*	Araceae
4.	Four leaf water clover	*Marsilea hirsute*	Marsileaceae
5.	Common reed	*Phragmites australis*	Poaceae
6.	Arrowhead	*Sagittaria subulata*	Alismataceae
7.	Cattail	*Typha latifolia*	Typhaceae
8.	Narrow leaf cattail	*Typha angustifolia*	Typhaceae
9.	Hydrilla	*Hydrilla verticillata*	Hydrocharitaceae
10.	Hornwort	*Ceratophyllum demersum*	Ceratophyllaceae
11.	Pond weed	*Potamogeton Perfoliatus*	Potamogetanaceae
12.	Parrot's feather	*Myriophyllum aquaticum*	Haloragaceae

S. No.	English Name	Scientific Name	Family
13.	Mosquito fern/ water fern	*Azolla filiculoides*	Salviniaceae
14	Crested floating heart	*Nymphoides cristata*	Menyanthaceae
15.	Water spangles	*Salvinia minima*	Salviniaceae
16.	Duckweed	*Wolffia arrhiza*	Lemnaceae
17.	Common duck weed	*Lemna minor*	Araceae
18.	Eel grass	*Vallisneria spiralis*	Hydrocharitaceae
19.	Pond weed	*Potamogeton pectinatus*	Potamogetonaceae
20.	Pond weed	*Potamogeton crispus*	Potamogetonaceae
21.	Hornwort	*Ceratophyllum demersum*	Ceratophylaceae
22.	Lotus	*Nelumbo lutea*	Nymphaceae
23.	Great duck weed	*Spirodela polyrhiza*	Lamnaceae
24.	Bulrushes	*Cyperus papyrus*	Cyperaceae
25	Indian pennywort	*Centella asiatica*	Umbelliferae
26.	Water hyssop	*Bacopa monnieri*	Scrophulariaceae
27.	Chinese water chestnut	*Trapa natans*	Trapaceae
28.	Blue speedwell	*Veronica anagallis*	Plantagonaceae
29.	Water primrose	*Ludwigia repens*	Onagraceae
30.	Water primrose	*Ludwigia inclinata*	Onagraceae
31.	Water primrose	*Ludwigia ovalis*	Onagraceae
32.	Water primrose	*Ludwigia palustris*	Onagraceae
33.	Water primrose	*Ludwigia Pantanal*	Onagraceae

English Name - Water Grass **Scientific Name** -*Echinochloa Colona* **Family**-Poaceae **Economics-** It is a valuable fodder and traditionally used in spleen and hemorrhage problems	**English Name-** Barnyard Grass **Scientific Name-***Echinochloa Crusgalli* **Family-** Poaceae

English Name – Love Grass **Scientific Name** –*Eragrostis Pilosa* **Family** - Poaceae **Economics** - The grass has some value as a forage and fodder. The grain is edible by humans.	**English Name** – Feather Love Grass **Scientific Name** – *Eragrostis Tenella* **Family-** Poaceae

English Name – Knot Grass **Scientific Name** – *Paspalum Distichum* **Family-** Poaceae	**English name** – Seashore Paspalum **Scientific Name** – *Paspalum Vaginatum* **Family-** Poaceae

English Name- Rough Bristle Foxtail **Scientific Name -** *Setaria Verticillata* **Family-**Poaceae	**English Name -** Green Foxtail **Scientific Name -** *Setaria Viridis* **Family-** Poaceae

English Name - Wrinkle Grass **Scientific Name –** *Ischaemum Rugosum* **Family –** Poaceae	**English Name-** Sandbur **Scientific Name –** *Cenchrus Biflorus* **Family-** Poaceae **Economics-** Cenchrus biflorus-seeds are used in Rajasthan and its Marwar region to make bread, either alone or mixed with bajra (millet)

English Name – Crow Foot Grass **Scientific Name** –*Dactyloctenium Aegyptium* **Family -** Poaceae	**English Name** – Crab Grass **Scientific Name**–*Digitaria Sanguinalis* **Family –** Poaceae

English Name - Jimson Weed

Scientific Name - *Datura Stramonium*

Family – Solanaceae

Economics- Ageratum conyzoides Medicinal ± leaves pounded to treat wounds. Popular with the Luo in Kenya as a haemostatic. Also known as a remedy for stomach pains.

Other Uses - A. hybridus can be used to make a red dye.

Datura stramonium Medicine ± In Zimbabwe leaves burned and smoke inhaled as a treatment for asthma. Leaves and seed contain the alkaloids hyoscine and atropine and are crushed and mixed with ghee to make an ointment for treatment of ringworm. Crushed leaves used as insect repellent. Other uses ± in Malawi it is believed that the leaves sprinkled around the house repel cockroaches. Datura spp. used as a rat poison in stored grain.

English Name - Smooth pigweed

Scientific Name - *Amaranthus Viridis*

Family - Amaranthaceae

English Name- Indian Goose Grass

Scientific Name - *Eleusine Indica*

Family- Poaceae

Economics- Eleusine indica Food-seed can be pounded into ¯our; eaten as a famine food in Zambia and Ethiopia. Other uses – straw used for bedding.

English Name- Red Sprangletop

Scientific Name - *Leptochloa Chinensis*

Family- Poaceae

English Name - Spiny pigweed

Scientific Name- *Amaranthus spinosus*

Family- Amaranthaceae

Economics- In the folk medicine of India, the ash of fruits of Amaranthus spinosus is used for jaundice. Water extracts from its roots and leaves have been used as a diuretic in Vietnam.

English Name - False Amaranth

Scientific Name - *Digera Arvensis*

Family- Amaranthaceae

English Name- False Amaranth **Scientific Name -** *Digera Muricata* **Family-** Amaranthaceae	**English Name –** Blistering Ammannia **Scientific Name –** *Ammannia Baccifera* **Family-** Lythraceae

English Name – Pink Node Flower **Scientific Name –** *Caesulia Axillaris* **Family -** Astereceae	**English Name –** Day Flower **Scientific Name –** *Commelina Benghalensis* **Family -** Commelinaceae

<table>
<tr><td></td><td></td></tr>
<tr><td>English Name – Spreading Day Flower
Scientific Name – Commelina Diffusa
Family - Commelinaceae
Economics - Commelina spp. Food - in times of famine leaves and young shoots used either fresh or boiled as a vegetable. Rhizome can also be eaten.
Medicine - C. africana used to treat leprosy, eye problems and colds.</td><td>English Name- Slender Day Flower
Scientific Name - Commelina Erecta
Family- Commelinaceae</td></tr>
<tr><td></td><td></td></tr>
<tr><td>English Nname- Wild jute
Scientific Name - Corchorus tridens
Family- Malvaceae
Corchorus spp. Food ± One of the most valuable local vegetables in hot, arid regions. Often cooked with potash. In Senegal the leaf is used with steamed millet to prepare `m'boum'. Eaten by women, children in many parts of Africa.
Medicinal ± C. olitorius used as a tonic and to treat toothache, stomach and bladder complaints.
Other uses - can be woven to make course cloth.</td><td>English Name – Petty spurge
Scientific Name – Euphorbia prostrate
Family- Euphorbiaceae
Economics- Euphorbia heterophylla Medicinal ± an infusion of leaves and root used as a remedy for headache.
Euphorbia hirta Medicinal ± Used to treat wounds by covering with fresh leaves. Common name of asthma weed but this may be because it is believed to induce asthma. The latex is sometimes used to treat eye complaints such as conjunctivitis.</td></tr>
</table>

English Name – Shrubby Spurge

Scientific Name – *Euphorbia Microphylla*

Family - Euphorbiaceae

English Name – Ground cherry

Scientific Name – *Physalis Minima*

Family- Solanaceae

English Name – Cockle Bur

Scientific Name – *Xanthium Strumarium*

Family – Astereceae

Economics - Xanthium strumarium is known as cang er in traditional Chinese medicine. Xanthium is also used to treat nasal and sinus congestion.

English Name- Indian turnsole/Devil weed

Scientific Name – *Heliotropium Indicum*

Family- Boraginaceae

English Name- Black Nightshade

Scientific Name – *Solanum Nigrum*

Family –Economics- Solanum nigrum Food - the unripe fruit are poisonous but in many parts of Africa the leaves are cooked as a relish.

Medicinal - Juice from the fruit has antibacterial and antifungal properties and is used to treat skin infections including ringworm

English Name – False Daisy

Scientific Name – *Eclipta Alba*

Family- Astereceae

English Name – Yellow Spider Flower

Scientific Name – *Cleome Viscosa*

Family- Brassicaceae

English Name – Purselane

Scientific Name – *Portulaca Oleracea*

Family- Portulaceae

Economics- Potulacca olerecea: Food - contains high levels of oxalic acid. May be eaten in salads and soups particularly in Mozambique.

Malawi. Large leaved types cooked as a vegetable.

Medicinal - as a snake bite remedy.

English Name – Desert horse purslane

Scientific Name – *Trianthema Portulacastrum*

Family- Aizoaceae

English Name-Puncture Vine

Scientific Name -*Tribulus Terrestris*

Family- Zygophyllaceae

English Name- Gulf leaf flower

Scientific Name - *Phyllanthus Niruri*

Family- Phyllanthaceae

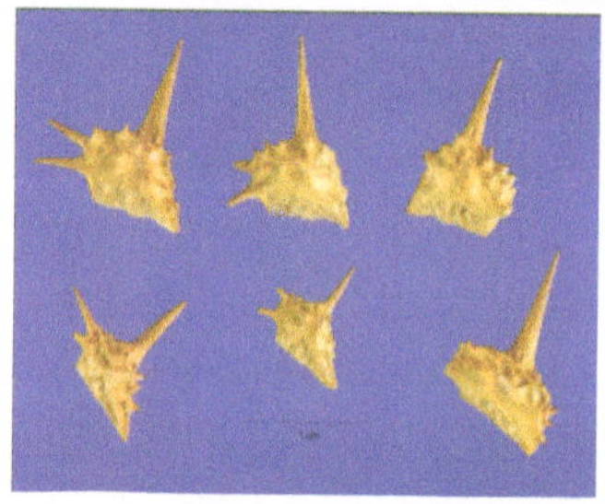

Burs of Tribulus Terrestris

English Name- Tiger Foot Morningglory

Scientific Name - *Ipomoea \ Ppestigridis*

Family- Convolvulaceae

English name- Blue Morning Glory

Scientific name - *Ipomoea Nil*

Family- Convolvulaceae

English Name - Coffee weed/ Sickle pod

Scientific Name - *Cassia Tora*

Family - Fabaceae

Economics- Cassia tora- It is used as a natural pesticide in organic farms, and as a powder commonly used in the pet food industry.

The seeds and leaves are used to treat skin disease and its seeds can be utilized as a laxative.

Since Cassia tora has an external germicide and antiparasitic character, it has been used for treating skin diseases such as leprosy, ringworm, itching and psoriasis and also for snakebites. Other medicinal provisions from plant parts include balm for arthritis using the leaves.

English Name- Creeping Indigo

Scientific Name - *Indigofera Linnaei*

Amily- *Fabaceae*

English Name- Wild Carrot Weed

Scientific Name - *Parthenium hysterophorus*

Family- Asteraceae

English Name - Tridex daisy

Scientific Name - *Tridax procumbens*

Family- Asteraceae

English Name - Grass Like Fimbry **Scientific Name -** *Fimbristylis Tenera* **Family -** Cyperaceae	**English Name -** Common Sedges **Scientific-** *Cyperus Difformis* **Family-** Cyperaceae

English Name - Yellow Berried Nightshade **Scientific Name -** *Solanum Xanthocarpum* **Family-** Solanaceae	**English Name-** Golden Daisy **Scientific Name -** *Vicoa Indica* **Family-** Astereceae

English Name- Worm Weed **Scientific Name -** *Artemisia scoparia* **Family-** *Asteraceae*	**English Name-** Burn Mouth Vine **Scientific Name -** *Rhyncosia Minima* **Family-** Fabaceae

English Name- Sun Spurge

Scientific Name - *Euphorbia Helioscopia*

Family- Euphorbiaceae

English Name- Chicken Eeed

Scientific Name- *Euphorbia Thymifolia*

Family- Euphorbiaceae

English Name- Common Spurge

Scientific Name - *Croton Sparsiflorus*

Family- Euphorbiaceae

English Name- Purselane

Scientific Name - *Portulaca Grandiflora*

Family- Portulaceae

English Name – Purselane

Scientific Name- *Portulaca Oleracea*

Family- Portulacaceae

Economics- Purslane has a high tolerance for salt toxicity, which makes it suitable for cropping in areas where irrigation is necessarily carried out with water having a high chloride-based salinity.

English Name- Pink Wood Sorrel

Scientific Name- *Oxalis Articulate*

Family- Oxalidaceae

English Name - Wild Buck Weed
Scientific Name - *Polygonum Glabrum*
Family - Polygonaceae

English Name- Small Knot Weed
Scientific Name - *Polygonum Plebeium*
Family- Polygonaceae

English Name- Shrubby Senna
Scientific Name - *Cassia Occidentalis*
Family- Fabaceae

English Name- Arrowhead
Scientific Name-*Sagittaria Sagittifolia*
Family- Alismataceae

English Name- Frogfruit
Scientific name - *Lippia Nodiflora*
Family- Verbenaceae

English Name - Hairy slitwort
Scientific Name - *Lindernia Ciliate*
Family- Scrofulariaceae

English Name - Winter Wild Oat
Scientific Name - *Avena Ludoviciana*
Family - Poaceae

English Name- Wild Oat
Scientific Name - *Avena Fatua*
Family- Poaceae

English Name- Sweet Grass
Scientific Name - *Poa Annua*
Family - Poaceae

English Name- Beard Grass
Scientific Name - *Polypogon Monspeliensis*
Family- Poaceae

English Name - Poison Rye Grass
Scientific Name - *Lolium Temulentum*
Family - Poaceae

English Name - Canary Grass
Scientific Name - *Phalaris minor*
Family - Poaceae
Economics- *Phalaris minor* - It is used as a fodder or forage for livestock and birdseed, but is poisonous to some mammals, and is a potential contaminant of seed crops. Popular mimicry weed of wheat.

<table>
<tr><td></td><td></td></tr>
<tr><td>English Name- Wild Onion
Scientific Name- Asphodelus Tenuifolius
Family - Liliaceae</td><td>English Name - Wild mustard
Scientific Name - Sisymbrium irio
Family - Brassicaceae</td></tr>
<tr><td></td><td></td></tr>
<tr><td>English Name -Wild Fenugreek
Scientific Name -Trigonella polycerata
Family-Fabaceae</td><td>English Name - Common Lambsquarter
Scientific Name - Chenopodium Album
Family - Chenopodiaceae
Economics- Chenopodium spp. Medicinal - C. ambosioides leaves applied to face to treat convulsions. Also powdered leaf mixed with oil and applied to skin to treat ringworm</td></tr>
</table>

English Name - Nettle Leaf

Scientific Name- *Chenopodium Murale*

Family - Chenopodiaceae

English Name - Stone Seed

Scientific Name - *Lithospermum Arvense*

Family - Boraginaceae

English Name - Green field-Speedwell

Scientific Name -*Veronica Agrestis*

English Name- Dock/Sorrel

Scientific Name - *Rumex Spinosus*

Family- Polygonaceae

English Name- Blue Daisy

Scientific Name - *Cichorium Intybus*

Family- Asteraceae

English Name- Wild Safflower

Scientific Name - *Carthamus Oxyacantha*

Family- Asteraceae

English Name- Maxican poppy

Scientific Name - *Argemone mexicana*

Family- Papaveraceae

Economics- Argemone mexicana Other uses - Seeds are narcotic and used to make traditional beer more intoxicating. Seeds have insecticidal properties. Can be poisonous to livestock.

English Name- Saw Thistle

Scientific Name - *Sonchus Arvensis*

Family - Asteraceae

English Name - Little Mellow

Scientific Name - *Malva Parviflora*

Family - Malvaceae

English Name- Blue Pimpernel

Scientific Name - *Anagallis Arvensis*

Family - Primulaceae

English Name- Purple Cud Weed

Scientific Name - *Gnaphalium purpureum*

Family- Asteraceae

English Name- Hairy vetch

Scientific Name - *Vicia Hirsute*

Family- Fabaceae

English Name- Vetch

Scientific Name - *Vicia Sativa*

Family- Fabaceae

English Name - Fumatory

Scientific Name - *Fumaria Parviflora*

Family - Fumariaceae

English Name- Garden cress

Scientific Name - *Coronopus Didymus*

Family- Brassicaceae

English Name- Corn Spurry

Scientific Name - *Spergula Arvensis*

Family- Caryophyllaceae

English Name- Cutleaf Evening primrose

Scientific Name - *Oenothera Laciniata*

Family- Onagraceae

English Name- Cursed Buttercup

Scientific Name- *Ranunculus Sceleratus*

Family- Ranunculaceae

English Name - Rough Seed Buttercup

Scientific Name - *Ranunculus* **Muricatus Family-** Ranunculaceae

English Name- Wild Dog Flower

Scientific Name - *Antirrhinum Orontium*

Family - Scrophulariaceae

English Name- Catch weed

Scientific Name - *Galium Aparine*

Family - Rubiaceae

English Name- Jhonson Grass

Scientific Name - *Sorghum Halepense*

Family - Poaceae

English Name- Tiger Grass **Scientific Name** - *Saccharum Spontaneum* **Family-** Poaceae	**English Name-** Bermuda Grass **Scientific Name** - *Cynodon Dactylon* **Family-** Poaceae **Economics-** Medicinal ± known in Malawi as a remedy for indigestion and heartburn.Other uses - pasture grass.

English Name- Nut Grass **Scientific Name** - *Cyperus Rotundus* **Family-** Cyperaceae **Economics-** Food ± Tubers are rich in starch and oil and it is a good source of phosphorous and iron, although protein content is relatively low, it is twice as high as in cassava. Tubers eaten raw or can be roasted to make`coffee', in Zambia and Zimbabwe used to make porridge	**English Name** - Field Bind Weed **Scientific Name** - *Convolvulus Arvensis* **Family** - Convolvulaceae

English Name- Blush Morning Glory
Scientific Name - *Ipomoea Carnea*
Family- Convolvulaceae

English Name- Prickly pear
Scientific Name - *Opuntia Dillenii*
Family- Cactaceae

English Bame- Canada thistle
Scientific Name - *Cirsium Arvense*
Family- Astereceae

English Name- Lantana
Scientific Name - *Lantana Camara*
Family - Verbinaceae

English Name- Goat Weed
Scientific Name - *Ageratum conyzoides*
Family- Astereceae
Economics- Medicinal ± leaves pounded to treat wounds. Popular with the Luo in Kenya as a haemostatic. Also known as a remedy for stomach pains.
Other uses - A. hybridus can be used to make a red dye.

English Name - Halfa Grass
Scientific Name - *Desmostachya Bipinnata*
Family - Poaceae

English Name- Congo Grass **Scientific Name -** *Imperata Cylindrical* **Family –** Poaceae **Economics- Medicinal -** known in Malawi as a treatment for digestive complaints	**English Name -** Marvel Grass **Scientific Name -** *Dichanthium Annulatum* **Family -** Poaceae

English Name – Spiderling **Scientific Name -** *Boerhavia Diffusa* **Family -** Nyctaginaceae	**English Name -** Spanish Needle **Scientific Name -** *Bidens Pilosa* **Family –** Asteraceae **Economics-** Food - leaves and shoots are edible. It is one of the weeds most widely used as a famine food due its abundance, rather than its taste which is rather aromatic. **Medicinal -** Leaves can be made into a poultice to treat wounds and the juice used to treat eye complaints. Roots and stem used to treat diarrhoea and abdominal pains.

English Name- Dallis Grass
Scientific Name - *Paspalum Dilatatum*
Family- Poaceae

English Name- Caesar's Weed
Scientific Name - *Urena Lobate*
Family- Malvaceae

English Name- Loranthus
Scientific Name - *Loranthus Micranthus*
Family- Loranthaceae

English Name - Dodder
Scientific Name - *Cuscuta Reflexa*
Family - Convolvulaceae

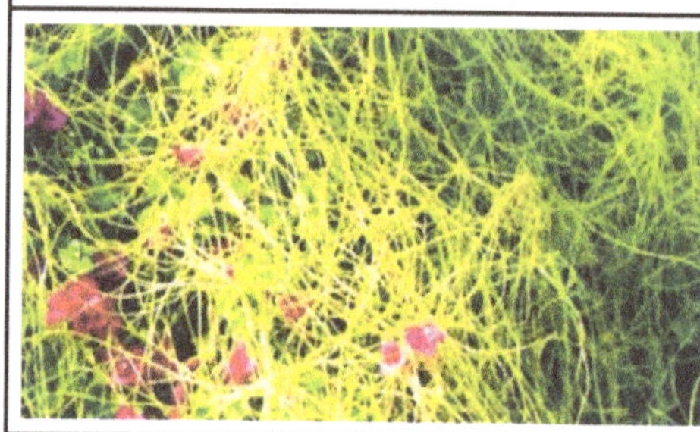

English Name- Dodder
Scientific Name - *Cuscuta Chinensis*
Family- Convolvulaceae

English Name - Dodder
Scientific Name - *Cuscuta Europaea*
Family - Convolvulaceae

English Name- Witch Weed
Scientific Name - *Striga Lutea*
Family- Scropulariaceae

English Name- Egyptian Broomrape
Scientific Name - *Orobanche aegyptiaca*
Family- Orabanchaceae

English Name- Broomrape
Scientific Name - *Orobanche Cernua*
Family – Orabanchaceae

English Name- Hemp broomrape
Scientific Name - *Orobanche Ramose*
Family - Orabanchaceae

English Name- Water Hyacinth
Scientific Name - *Eichhornia Crassipes*
Family- Pontederiaceae

English name - White Water Lily
Scientific Name - *Nymphaea Alba*
Family - Nymphaeaceae

English Name- Water Lettuce
Scientific Name- *Pistia Stratiotes*
Family- Araceae

English Name- Four Leaf Water Clover
Scientific Name - *Marsilea Hirsute*
Family- Marsileaceae

English Name- Common Reed
Scientific Name - *Phragmites Australis*
Family- Poaceae

English Name- Arrowhead
Scientific Name - *Sagittaria Subulata*
Family- Alismataceae

English Name- Cattail
Scientific Name - *Typha Latifolia*
Family- Typhaceae

English Name - Narrow Leaf Cattail
Scientific Name - *Typha Angustifolia*
Family- Typhaceae

English Name - Parrot's Feather
Scientific Name - *Myriophyllum Aquaticum*
Family - Haloragaceae

English Name - Mosquito Fern/ Water Fern
Scientific Name - *Azolla Filiculoides*
Family - Salviniaceae

English Name - Crested Floating Heart
Scientific Name - *Nymphoides Cristata*
Family - Menyanthaceae

English Name - Water Spangles
Scientific Name - *Salvinia Minima*
Family - Salviniaceae

English Name - Duckweed
Scientific name - *Wolffia Arrhiza*
Family - Lemnaceae

English Name- Pond Weed
Scientific Name - *Potamogeton Pectinatus*
Family- Potamogetonaceae

English Name- Pond Weed
Scientific Name - *Potamogeton Crispus*
Family- Potamogetonaceae

English Name - Hornwort
Scientific Name - *Ceratophyllum demersum*
Family- Ceratophylaceae

English Name- Lotus
Scientific Name - *Nelumbo Lutea*
Family- Nymphaceae

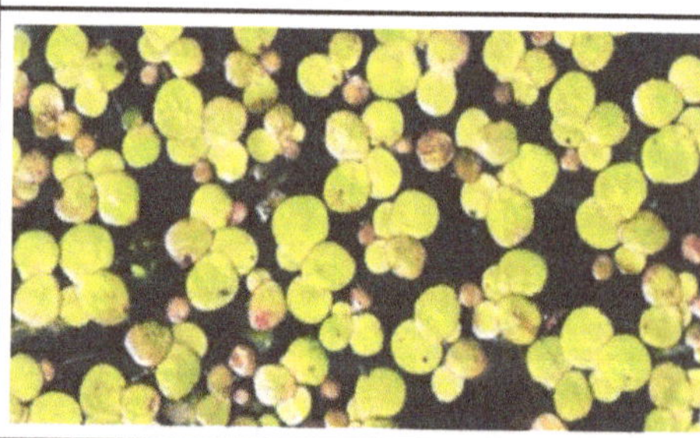

English Name - Great Duck Weed
Scientific Name - *Spirodela Polyrhiza*
Family- Lamnaceae

English Name- Bullrushes
Scientific Name - *Spirodela Polyrhiza*
Family- Cyperaceae

English Name- Chinese Water Chestnut
Scientific Name - *Trapa Natans*
Family- Trapaceae

English Name- Blue Speedwell
Scientific Name - *Veronica Anagallis*
Family- Plantagonaceae

English Name - Water Primrose
Scientific Name - *Ludwigia Repens*
Family - Onagraceae

English Name- Water Primrose
Scientific Name - *Ludwigia Inclinata*
Family- Onagraceae

English Name - Pignut
Scientific Name - *Hyptis Suaveolens*
Family- Lamiaceae

English Name - Water Primrose

Scientific Name - *Ludwigia Ovalis*

Family- Onagraceae

English Name- Water Primrose

Scientific Name - *Ludwigia Palustris*

Family- Onagraceae

English Name- Water Primrose

Scientific Name - *Ludwigia pantanal*

Family-Onagraceae

English Name- Naregamia

Scientific Name - *Naregamia Alata*

Family- Miliaceae

English Name - Rough Bryony

Scientific Name - *Mukia Maderaspatana*

Family- Cucurbitaceae

English Name- Elephant Foot

Scientific Name - *Elephantopus Scaber*

Family- Asteraceae

English Name- Indian iquorice

Scientific Name - *Abrus Precatorius*

Family - Fabaceae

English Name- Indian Copper Leaf

Scientific Name - *Acalypha Indica*

Family - Euphorbiaceae

English Name - Javanese Wool Plant

Scientific Name - *Aerva Lanata*

Family - Amaranthaceae

English Name- Indian Joint Vetch

Scientific Name - *Aeschynomene Indica*

Family- Fabaceae

English Name - Life Plant

Scientific Name - *Biophytum Sensitivum*

Family - Oxalidaceae

English Name- Blainvillea

Scientific Name - *Blainvillea Rhomboidea*

Family- Asteraceae

English Name- Chamaelea

Scientific Name - *Sebastiana Chamaelea*

Family- Euphorbiaceae

English Name- Touch Me Not
Scientific Name - *Mimosa Pudica*
Family- Fabaceae

English Name - Giant Sensitive Weed
Scientific Name - *Mimosa Invisa*
Family – Fabaceae

English Name- Sensitive Plant
Scientific Name - *Mimosa Diplotricha*
Family - Fabaceae

English Name- Rose Periwinkle
Scientific Name - *Vinka Rosea*
Family - Apocynaceae

English Name- Sweet Broom Weed
Scientific Name - Scoparia Dulcis
Family- Scrophulariaceae

Index

Printed at Repro India Ltd.